Lecture Notes in Statistics 189

Edited by P. Bickel, P. Diggle, S. Fienberg, U. Gather,
I. Olkin, S. Zeger

D1539116

Bill Thompson

The Nature of
Statistical Evidence

 Springer

Bill Thompson
Department of Statistics
University of Missouri-Columbia
Columbia, Mo 65211
e-mail: pickensT@missouri.edu

Library of Congress Control Number: 2006938160

ISBN-10: 0-387-40050-8   e-ISBN-10: 0-387-40054-0
ISBN-13: 978-0-387-40050-1   e-ISBN-13: 978-0-387-40054-9

Printed on acid-free paper.

9 8 7 6 5 4 3 2 1

springer.com

Contents

III. STATISTICAL MODELS OF INDUCTION

Preface

Our motivation for writing this book was a dissatisfaction with the many books with titles like *Foundations of Statistics*. These books provide a needed description of the subject with examples of various statistical methods; but they do not satisfy the discerning reader because they do not explain why certain conclusions may be drawn from certain data and they do not discuss how statistics, the subject, meshes with the scientific process. We naively set out to fill in these gaps, but the situation is not so simple. What is desired is a tool—the one true statistics—which can be applied to data with certainty to justify conclusions, but what is available, in fact, are various competing theories of statistics.

The careful reader may correctly point out the relation of this manuscript to postmodern epistemology, a philosophy which emphasizes that all human knowing—in fields as diverse as religion and science—is culture dependent and that therefore "truth" is not absolute. Take, for example, the "truths" of Christianity and Islam—wars have been fought over their differences. In the field of statistics there are, for example, Bayesians and sampling theorists. Carson (2003) states, "Surely it is better, postmoderns tell us, to encourage insights that flow from many different perspectives . . . " In statistics, postmodernism is manifest at two levels. First, the findings of statistical applications are not about the true state of nature but about models of experiments on nature. Second, the theorems and concepts of statistical theory are not even about models of nature, they are about whatever is postulated by thought, and cannot be checked by experiment. What experiment on nature could check the efficacy of, say, the likelihood principle?

The purpose of this book is to discuss whether statistical methods make sense. That is a fair question, at the heart of the statistician–client relationship, but put so boldly it may arouse anger. Still, we statisticians could be wrong and we need a better answer than just shouting a little louder. To avoid a duel, we prejudge the issue and ask the narrower question: "In what sense do statistical methods provide scientific evidence?"

Shewhart (1939, p. 48) writes—

. . . the statistician of the future (in applying statistical theories to quality control) . . . must go further than is customarily recognized in the current literature in that he must provide operationally verifiable meanings for his statistical terms such as random variable, accuracy, precision, true value, probability, degree of rational belief, and the like.

Shewhart's words will serve as an introduction to the present writing. Little progress has been made toward providing meanings for statistical terms; statisticians have not considered that their theories and methods have no practical meaning until their basic concepts are interpreted. Neither data nor mathematics speak for themselves; they have to be interpreted and the interpretation defended. The present volume begins the task of providing interpretations and explanations of several theories of statistical evidence.

Another good introduction is Efron (2004).

Have we achieved a true *Foundations of Statistics?* We have made the link with one widely accepted view of science and we have explained the senses in which Bayesian statistics and *p*-values allow us to draw conclusions. While our treatment is not complete, we have written a needed text and reference book on the foundations of statistics. We have not settled the controversy between Bayesians inference and sampling theory.

In some ways our story begins with deductive logic—a subject with which most readers will be familiar. For these reasons we provide an appendix on logic which the reader may use accordingly if the need arises.

The author wishes to thank the University of Missouri, Department of Statistics, particularly Tracy Pickens, for much help in production of this manuscript. Thanks are also due to the editor, John Kimmel, and several anonymous readers who substantially improved the exposition.

1

Overview

We begin by clarifying terms. *Evidence* is grounds for belief—an imprecise concept. There must be many valid reasons for believing and hence many ways of making the concept of evidence precise. Most of our beliefs are held because mother—or someone else we trust—told us so. The law trusts sworn testimony. Scientific and statistical evidence are other, different grounds for conclusion, supposedly particularly reliable kinds.

Scientific thinking is an approach to knowing, which emphasizes system and objectivity. Its success can be seen everywhere, but, while there is a 500-year-old scientific spirit, there still is no agreed-upon scientific method. The stated goals of scientists vary; they describe, explain, summarize, and/or predict. The explanatory power of science satisfies the human need to understand, description records the facts, and summarization makes the whole comprehensible. Prediction facilitates design, arranging things to achieve a desired end.

Most treatments of science are implicitly restricted to experimental science. An experiment is a text describing the recipe for performance. An experimental science consists of statements about experiments; these statements are not "beliefs" but Tukey (1960, pp. 425–6) "conclusions."

A conclusion is a statement which is to be accepted as applicable to the conditions of an experiment or observation unless and until unusually strong evidence to the contrary arises.

But science is not—as it was treated prior to Kuhn (1962)—an automatic and impersonal process of logic and experimentation; truth does not simply leap from the test tubes. Experiments have to be interpreted and conclusions defended. Deduction and experimentation are the permissible arguments, but the short-term test of a conclusion is peer review.

Regularities do exist in nature and conclusions can be checked; some conclusions are just better able to withstand experimental test than others. Scientists who hold these more fit (in the evolutionary sense) views are more successful in solving problems and convincing colleagues of the correctness of their solutions. Theories evolve in the manner in which genetic traits evolve: primarily through the professional survival and evolution of the scientists who carry them. Thus, it is not the personal considerations of what to believe or how to act which are central

1

to scientific process, but rather the social matter of how one scientist convinces another of his conclusions.

Example 1.1 The law is an example of one place where the nature of scientific and statistical evidence becomes crucial.

Suspicion is the basic legal attitude of the courts toward scientific evidence; courts

... have sometimes been skeptical of scientists' claims of the of the virtual infallibility of scientific techniques.

Imwinkelried (1989, p. 90)

Consequently, most courts have maintained that scientific evidence is inadmissible unless it is established

... that the theory and instrument have been generally accepted within the relevant scientific circles.

Imwinkelried (1989, p. 91)

The United States Supreme Court decision in Daubert v. Merrell-Dow Pharmaceuticals, Inc. 113 S. Ct. 2786, 125 L. Ed. 2d469 (1993) marked a recent change in the legal attitude toward science. The Daubert decision replaces the "generally accepted" requirement by "scientific knowledge." To be admissible, testimony must be "scientific knowledge" according to the Pearsonian view that it is the product of sound scientific methodology.

The Court defined scientific methodology as the process of formulating hypotheses and then conducting experiments to prove or falsify the hypothesis. According to the majority opinion in Daubert, in deciding whether a proposition rests on sound scientific methodology, the trial judge should consider such factors as whether the proposition is testable, whether it has been tested, the validity rate attained in any tests, whether the research has been peer reviewed, and whether the findings are generally accepted.

Imwinkelried (1989, p. 91)

Statistical evidence is no more readily accepted by the courts than other scientific evidence.

Using ordinary statistical inference as proof of commonplace events or conditions has had a rocky history in the judicial system. On the civil side, in what are often called "blue bus" cases, courts rebel at arguments that defendant's bus or tire caused the injury simply because the defendant is responsible for most of the buses or tires that might have caused the injury. And in the landmark criminal decision in the Collins case, where eyewitness testimony indicated that a black man and a blond woman mugged a woman in an alley in Los Angeles, the California Supreme Court overturned a conviction after a trial in which the prosecutor argued that various human qualities common to the culprits and the defendants are so rare that statistical probabilities heavily favored the proposition that defendants were the culprits.

Mueller and Kirkpatrick (1999, p. 953)

Interpretations of probability and explanations of statistical inference will be the topics of the remainder of this book. Details of the Collins case "were that the man had a beard and a mustache, the woman had a ponytail, the two comprised an interracial couple, and they drove off in a partly yellow convertible." The prosecutor suggested that the probability of finding all qualities together should be obtained by multiplying the probabilities of individual qualities to obtain a probability of 1 in 12 million. The rationale given by Mueller and Kirkpatrick (1999, p. 954) for reversal is that "statistical inference . . . should not be allowed when there is no proof supporting the suggested frequencies or probabilities." In Collins, for example, no empirical justification for multiplying probabilities was given, and

> There was no proof that one car in ten is partly yellow or that one couple in a thousand is "interracial." These were made-up numbers, and the point is not so much that they are too high or too low, but that they purport to add precision to common insights that resist such refinement. The tactic adds a false patina of science to the everyday process of interpretation and evaluation, . . .

We discuss this example further in Section 14.5.

We now turn to a general discussion of statistics. Immanuel Kant pointed out in the middle of the eighteenth century that mind is an active organ which processes experience into organized thought. There is rather little knowledge in a 2-foot stack of computer printout. It only becomes knowledge after its essential message has been highlighted by a process of organization. This is the function which statistical methods perform.

While it won't quite serve as a definition, that aspect of statistics which we consider is concerned with models of inductive rationality. We will be concerned with rational reasoning processes, called statistical methods, which transform statements about the outcome of particular experiments into general statements about how similar experiments may be expected to turn out. As such, statistical methods have aims similar to the process of experimental science. But statistics is not itself an experimental science, it consists of models of how to do experimental science. Statistical theory is a logical—mostly mathematical—discipline; its findings are not subject to experimental test. The primary sense in which statistical theory is a science is that it guides and explains statistical methods. A sharpened statement of the purpose of this book is to provide explanations of the senses in which some statistical methods provide scientific evidence.

We embed our discussion of statistical evidence into modern ideas about mathematics and scientific modeling. First, the axioms of statistical inference are like Euclid's axioms in that they are formally "empty" statements with no necessary implication for inference or space, respectively. But unlike Euclid's axioms, the axioms of inference cannot be checked against the world. The nature of a statistical theory is therefore that its truth is conditional on assumptions which cannot be verified. Secondly, we embed the discussion of statistical inference into modern ideas of scientific modeling; an inferential model depends on prevalent paradigms and background information. Therefore, findings are not about "true values" but parameters which best fit the target population.

Different laboratories applying the same experimental text to the same "thing" are different experiments requiring different parameterizations. The crucial issue concerning experimentation is *whether* results are reproducible by different laboratories. Birnbaum (1962) adopts an evidential approach to statistics, and others have followed. They write $E_v(E, y)$ for the evidential meaning of obtaining data y as the outcome of experiment E, but $E_v(E, y)$ does not exist. Rather we should ask about $E_v(E, T, y)$, the evidential meaning of observing y as outcome of E in the light of theory T, where T consists of assumptions about (i) the inferential logic being used, for example, Bayesian or p-values; and (ii) the target population being observed.

This book is not a complete discussion of statistical foundations. We focus on two models of statistical inference: Bayesian statistics and p-values.

Rubin (1984) describes inference to be Bayesian, if known as well as unknown quantities are treated as random variables—knowns having been observed but unknowns unobserved—and conclusions are drawn about unknowns by calculating their conditional distribution given knowns from a specified joint distribution. Thus, if Bayesian statistics is to explain why certain conclusions follow from certain data, we need an explanation of why the relevant quantities may be considered random. There are several appealing explanations.

One explanation of randomness, based on deFinetti's theorem, is that we are fitting a sequence of performances of an experiment by a class of exchangeable random variables. This explanation is not helpful in choosing a prior distribution from which to start and it does not yield inferential insight for independent and identically distributed random variables.

A second explanation is that we are quantifying the personal degree of belief of an economic man as evidenced by his betting behavior. We like these explanations of Bayesian inference. But economic Bayesian inference will not be useful in helping to resolve disagreements among the members of a group, where personal probability methods don't apply, since different members will have different prior beliefs.

Our second inference procedure of focus is significance testing, by which we here mean the theory of p-values as evidence. We generalize the notion of p-value to obtain a system for assessing evidence in favor of a hypothesis. It is not quite a quantification in that evidence is a pair of numbers (the p-value and the p-value with null and alternative interchanged) with evidence for the alternative being claimed when the first number is small and the second is at least moderately large. Traditional significance tests present p-values as a measure of evidence against a theory. This usage is rarely called for since scientists usually wish to accept theories (for the time being) not just good or bad for this purpose; their efficacy depends on specifics. We find that a single p-value does not measure evidence for a simple hypothesis relative to a simple alternative, but consideration of both p-values leads to a satisfactory theory. This consideration does not, in general, extend to composite hypotheses since there, best evidence calls for optimization of a bivariate objective function. But in some cases, notably one-sided tests for the exponential family, the optimization can be solved, and a single p-value does provide an appealing measure of best evidence for a hypothesis.

Our overall conclusion is that there are at least two kinds of statistical evidence, each of which lends a different explanatory insight, and neither of which is perfect. This essay describes the kinds of insight contributed and the circumstances for which each is appropriate. This chapter summarizes our findings. Supporting references and argument are to be found in the body of the book, but in a different order. It was convenient to group the discussion into three parts: related background material, probability, and inference.

I

The Context

2

Mathematics and Its Applications

Mathematical arguments are fundamentally incapable of proving physical facts.

Cramér (1946, p. 146)

2.1. Axiomatic Method and Mathematics

It is a historical fact that geometry was crucial in the development of the modern view of mathematics and the axiomatic method. David Hilbert (1862–1943) judged that the invention of non-Euclidean geometry was "the most suggestive and notable achievement of the last century," a very strong statement, considering the advance of science during the period. Hilbert meant that the concepts of truth and knowledge, and how to discover truth and acquire knowledge, all this was changed by the invention of non-Euclidean geometry. This chapter reviews some of that historical development.

In his *Elements* (330–320 B.C.) Euclid writes,

Let the following be postulated:

  I. To draw a straight line from any point to any point.
 II. To produce a finite straight line continuously in a straight line.
 III. To describe a circle with any center and distance.
 IV. That all right angles are equal to one another.
  V. That, if a straight line falling on two straight lines makes the interior angles on the same side less than two right angles, the two straight lines, if produced indefinitely, meet on that side on which are the angles less than the two right angles.

Most current work replaces Euclid's fifth postulate by the equivalent Playfair postulate: through a given point, not on a given line, exactly one parallel can be drawn to the given line. The reader will be familiar with the modern way of using Euclid's postulates to construct an extensive and worthwhile theory.

To illustrate the process, Wilder (1983, pp. 10–16) considers a simpler axiomatic system with just two undefined terms: point and line.

Axiom 1. Every line is a collection of points.

Axiom 2. There exist at least two points.

Axiom 3. If p and q are distinct points, then there exists one and only one line containing p and q.

Axiom 4. If L is a line, then there exists a point not on L.

Axiom 5. If L is a line, and p is a point not on L, then there exists one and only one line containing p that is parallel to L.

Wilder investigates the properties of this system by proving several theorems. A single illustration of the process will suffice for our purpose.

Theorem 2.1 Every point is on at least two distinct lines.

Proof

	Statement	Justification
i.	Consider any point p.	Axiom 2
ii.	There is a point q distinct from p.	Axiom 2
iii.	There is a line L containing p and q.	Axiom 3
iv.	There is a point r not on L.	Axiom 4
v.	There is a line K containing p and r.	Axiom 3
vi.	r is on K but not on L.	iv and v
vii.	L and K are distinct.	Axiom 1 and vi
viii.	p is on L and K.	iii and v

The important thing to notice about the above proof is that *only* axioms and logic are offered as justification. In particular we do not rely on geometric sketches nor on our experience of the space we live in. The reason for this will now appear. Wilder's axiom system reminds us of our high-school study of Euclidean geometry, but we have a different concept in mind. We are thinking of an agricultural field trial to increase crop yields. Four fertilizer treatments are to be applied to blocks of land, each consisting of two plots. The points of the system are the treatments denoted by {a,b,c,d}, the lines are the six blocks of land labeled by the pairs of treatments applied to their subplots: {a,b}, {a,c} {a,d}, {b,c},{b,d}, {c,d}. Note that the axioms are satisfied—two lines being parallel meaning that they do not contain a common point. Here, point is interpreted as "treatment" and "line" is interpreted as "block." The agricultural field trail is an interpretation of the axiom system. The same axioms may serve for two, in fact many, interpretations.

It is not sufficiently appreciated that a link is needed between mathematics and methods. Mathematics is not about the world until it is interpreted and then it is only about models of the world (Eves 1960, p. 149). No contradiction is introduced by either interpreting the same theory in different ways or by modeling the same concept by different theories. This is the message of the observation of Cramér with which we began this chapter.

In general, a primitive undefined term is said to be **interpreted** when a meaning is assigned to it and when all such terms are interpreted we have an **interpretation** of the axiomatic system. It makes no sense to ask which is the correct interpretation of an axiom system. This is a primary strength of the axiomatic method; we can use it to organize and structure our thoughts and knowledge by simultaneously and economically treating all interpretations of an axiom system. It is also a weakness in that failure to define or interpret terms leads to much confusion about the implications of theory for application.

The axiomatic method introduces primitive terms (such as point and line) and propositions concerning these terms, called axioms. The primitive terms and axioms taken together are called the axiom system $\sum$. The primitive terms are left completely undefined: they are to be the subject of subsequent investigation. To emphasize that propositions are to be phrase-able in the language of the system $\sum$, they are called $\sum$-**statements**. Logic is applied to the axioms in order to derive theorems. The **theorems** of any axiomatic system are of the form $\sum \Rightarrow q$, where q is a $\sum$ statement. The logic used to prove theorems is usually the propositional calculus, informally applied. In the course of the investigation it is often convenient to define further terms. A definition is a characterization of the thing defined.

We might ask if it isn't a little foolish to discuss, at great length, primitive terms which are not defined. Why not define them so that we can agree on what we are talking about? The answer seems to be that definition of all terms isn't possible without circularity. Thus the dictionary defines truth as a verified fact and fact as a thing which is really true. Further, we might ask why the axioms should be accepted without proof. It is because "you don't get something for nothing." If nothing is assumed, then no theorems can be proved. The axiomatic approach is to agree conditionally on some basic principles and see where they lead, to ask "What if . . . ?" As a result of the ensuing investigation, altered basic principles can then be contemplated.

An axiom system is **consistent** if it does not imply any contradictory theorems. Only consistent systems are of interest since any $\sum$-statement, q, is a theorem of $\sum$ if $\sum$ is inconsistent. To see this, suppose that $\sum \Rightarrow p$ and $\sum \Rightarrow \sim p$. Then, since $\sim p \Rightarrow (p \rightarrow q)$, we have that $\sum \Rightarrow p \wedge (p \rightarrow q) \Rightarrow q$.

But it is not possible to prove the consistency of $\sum$ as a consequence of $\sum$. For suppose that c is the statement that $\sum$ is consistent. A truth table shows that $\sum \Rightarrow c$ is equivalent to $\sum \wedge \sim c \rightarrow c$. Thus if on the basis of the axioms, we can show that the axioms are consistent then we will also have shown that if the axioms are true but contradictory then they are consistent. It is absurd that we should be able to prove this last statement.

Consistency of $\sum$ is essential but not provable. How then might we check for consistency? The answer requires the concept of interpretation. The working mathematician adopts the test that if $\sum$ admits an interpretation I then $\sum$ is consistent. For suppose $\sum \Rightarrow p$ and $\sum \Rightarrow \sim p$. We have $| \sum, I | = 1$ and hence $|p, I| = 1$ and $| \sim p, I | = 1$ denying the Aristotelian law of contradiction. Wilder (1983, p. 27) refers to this bit of informal reasoning as "basic principles of applied logic."

Bertrand Russell's famous definition of mathematics as "the subject in which we never know what we are talking about, nor whether what we are saying is true" perhaps suggests that pure mathematics is axiomatic method without regard to the validity of the primitive terms and the axioms. Recall that the primitive terms of an axiom system are undefined, axioms are set down without regard to their truth and the conclusions of the theorems may not be true since their hypotheses may be false.

But not all mathematicians agree with Russell; it has been remarked that it would be difficult to find two mathematicians who agree on the same definition. In desperation, mathematics has been defined to be "what mathematicians do when they are doing mathematics." The suggestion is not as facetious as it sounds; it indicates that mathematics is a cultural group activity, which is true.

Constructing and investigating axiomatic systems is certainly one of the things which some mathematicians do in a professional capacity. Other mathematicians find the axiomatic method too confining and do not take it seriously. A casual examination of the books on the author's shelf indicates that many mathematical texts and subjects are not developed from a single set of axioms, but the hypotheses of the theorems are altered as necessary to obtain results.

Well, when doing mathematics, mathematicians prove theorems and construct counterexamples. A **counterexample** to a conjectured theorem consists of exhibiting a logical possibility where the hypothesis is true but the conclusion is false. Mathematicians measure their day-to-day progress in terms of whether or not a theorem has been proved. Counterexamples seem to be equally important but for some reason, they are interpreted as indicating lack of progress. So, as far as it goes, perhaps mathematics is proving theorems. A theorem is a true statement of implication, of the form $p \Rightarrow q$; p is called the **hypothesis** of the theorem and q its **conclusion**. Recall that implication means truth preservation or that $p \to q$ is a tautology.

A proof of the theorem $p \Rightarrow q$ is an argument which shows $p \to q$ to be a tautology. Equivalence of the statements $p \to q$, $\sim q \to \sim p$, $(p \wedge \sim q) \to \sim p$, $(p \wedge \sim q) \to q$ and $(p \wedge \sim q) \to c$, where c is any contradiction provides several other important ways of proving $p \Rightarrow q$. The second and last ways are called "the indirect method of proof" and "reductio ad absurdum," respectively.

2.2. Propositional Calculus as an Axiomatic System*

Lukasiewicz (1929) takes $\sim$ and $\to$ as primitive and defines $p \vee q = \sim p \to q$ and $p \wedge q = \sim(p \to \sim q)$. He then assumes that the three statements (i) $(p \vee p) \to p$, (ii) $p \to (p \vee q)$ and (iii) $(p \to q) \to [(q \to r) \to (p \to r)]$ are tautologies and shows that all of the tautologies of the propositional calculus, as we have outlined it, can be derived from these three. We say that the propositional calculus is axiomatized by the primitive terms $\sim$ and $\to$ and the three assumed tautologies. The rules of derivation to be used are substitution and *modus ponens*. Substitution authorizes the replacement of any simple proposition by a compound proposition in an axiom or previously proved formula. *Modus ponens* (in other contexts called affirming the antecedent) is the rule that given p and $p \to q$ we may infer q. We illustrate the process of derivation with two examples. First, substituting $\sim p$ for p in statement ii we obtain

---

* Sections marked with an asterisk may be omitted.

~p→ (p→q), a formula which has been called a paradox of the conditional since it implies that a false proposition is sufficient for every proposition. In fact, if p is false then ~p is true so that from ~p→ (p→q) we obtain p→q by modus ponens. Second, assuming p→q, we have from statement iii by *modus ponens* that (q→r)→ (p→r). From this formula, assuming in addition q→r we obtain p→r by *modus ponens*. This is the transitive result that from p→q and q→r we may infer p→r.

Lukasiewicz's axioms may be verbalized as follows:

i.   if p is true or p is true than p is true
ii.   if p is true then p or q is true
iii. if p is sufficient for q then r or ~q is sufficient for r or ~p.

It is interesting that none of Aristotle's "basic laws of thought" are included among Lukasiewicz's axioms.

2.3. The Euclidean Geometry Paradigm[1]

The quote of Cramér, with which we began this chapter, might appear common-place except that for 2,000 years it was thought to be otherwise. What might be called the Euclidean geometry paradigm for learning about the world is the fol-lowing. There are intuitive entities such as point and line; one looks around till one finds "self-evident truths" concerning these entities, Euclid's postulates, for example. Logic is then used to derive, from the basic truths, more interesting and more complicated truths about the entities being studied. Rene Descartes in his *Discourse on Method* (after 1600) described the situation succinctly as follows:

The long chain of simple and easy reasoning by means of which geometers are accustomed to reach the conclusions of their most difficult demonstrations, has led me to imagine that all things, to the knowledge of which man is competent, are mutually connected in the same way, and there is nothing so far removed from us as to be beyond our reach or so hidden that we cannot discover it, provided only we abstain from accepting the false for true, and always preserve in our thoughts the order necessary for the deduction of one truth from another.

Euclidean geometry was conceived to be the study of the actual physical space in which we live, and its theorems were thought to be discovered truths about the world. Further, Euclidean geometry was the prototype of how other disciplines should proceed in their development.

There had been early indications of the inadequacy of the paradigm, but minds were not ready to receive them. That the world was flat and that the sun revolved about the earth were "self-evident truths"; one merely had to observe to see that they were true. By the sixteenth century both of these views had given way to superior theories.

---

[1] The Euclidean geometry paradigm is often called classical rationalism.

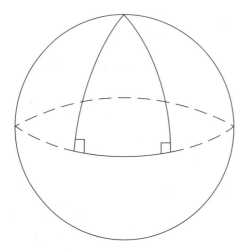

FIGURE 2.1. The sum of interior angles of a triangle exceeds two right angles.

Much later Einstein's theory of relativity would make the paradigm clearly untenable. Newtonian mechanics was and is a prime example of science. But if it has to be interpreted literally as what is going on in nature, then we must say that Einstein showed Newton to be wrong. For how can our one world be two different ways? But Newton wasn't wrong. Euclid was wrong—geometry is not about physical space, and scientific theory isn't about nature, only models of nature.

But it was the invention of non-Euclidean geometry which first and finally laid the paradigm to rest. If Euclid's second postulate is relaxed and the fifth changed so that every pair of lines meet, then one obtains elliptic geometry, useful in global navigation where the lines are great circles. An important consideration is to note that the properties of elliptic geometry are realized on a sphere (merging antipodal points) in Euclidean three-space and so any inconsistency in elliptic geometry is also an inconsistency in Euclidean geometry. Euclidean geometry was no more valid or correct than non-Euclidean. But the truth of a proposition depends on which geometry is postulated. In Euclidean geometry the sum of the interior angles of a triangle equals two right angles. But in elliptic geometry this sum exceeds two right angles. For example, consider the triangle of Figure 2.1, formed by the equator and two great circles through the poles.

In summary, equally valid geometries existed side by side, but their true propositions were in some instances mutually contradictory. It was clearly inappropriate to search for "the one true geometry." For the purpose of surveying, Euclidean geometry is the proper instrument whereas elliptic geometry is more useful for navigating a flight from New York to Tokyo. The adequacy of Euclidean geometry depends on the purpose to which it is put. How could this be if geometry were about physical space? The world can be only one way.

As a result, the geometry of Euclid ceased to be accepted as a paradigm for mathematics and scientific method at about the end of the nineteenth century. How can the axioms, and in particular the axiom of parallels, be self-evident if there are other interesting and productive but contradictory possibilities? C.S. Pierce wrote in 1891,

Geometry suggested the idea of a demonstrative system of absolutely certain philosophical principles, and the ideas of the metaphysicians have at all times been in large part drawn from mathematics. The metaphysical axioms are imitations of the geometrical axioms; and now the latter have been thrown overboard, without doubt the former will be sent after them.

Euclid, that well-meaning genius, misled us for 2000 years until the development of non-Euclidean geometries showed that geometry is not about physical space. Morris Kline (1980) in his book *Mathematics: The Loss of Certainty* pounds nail after nail into Euclid's coffin. The message of Kline's book is outdated by 100 years, but such ideas die hard. The Euclidean geometry paradigm has penetrated to the core of Western thought and it may be another 2000 years before "self-evident truths" disappear from the high-school curriculum. Self-evident truths are dead but not buried in matters of science, but they are very much alive where ideologies are the issue: "We hold these truths to be self-evident. That all men are created equal...." Getting ahead of our story, only when statistical inference is treated as an ideology are its basic premises "self-evident truths."

2.4. Formal Science and Applied Mathematics

The question of the source of the axioms of a system $\sum$ is especially important for our purpose. We have seen that they are not "self-evident truths." We can conceive of setting up an axiom system involving nonsense syllables just for the fun of it, but statements involving nonsense syllables do not readily suggest themselves and it is difficult to judge whether several such statements would be consistent. Most axiom systems are motivated by some primitive concept (as physical space motivates Euclidean geometry). The axioms are statements which seem to hold for the concept; this is the only sense in which axioms may be called true.

Feather (1959, p. 127) says of Newtons laws,

...we shall merely state the laws of motion as he formulated them, for however much we may work over the experimental results beforehand, we cannot deduce the laws by rigorous argument. There is much intuition in them, even some obscurity, and they stand to be tested in the light of subsequent experience, not to be passively accepted as established on the basis of previous knowledge-as indeed do all the laws of physics.

Normally the motivating concept precedes $\sum$, but once we have $\sum$ it often proves interesting to consider the consequence of altering its features. We then have an axiomatic system in search of an interpretation; it is surprising how often we find one.

Of course, axiomatic systems are constructed not only within mathematics proper, but also in mechanics and economics, for example. **Formal science** is axiomatic method done when we have some primitive concept in mind which we wish to investigate. A formal science will always be consistent since the primitive concept motivating $\sum$ will serve as an interpretation of $\sum$.

A desirable criterion in choosing the axioms and undefined terms of a formal science is simplicity or neatness. We may conceive of a totality, T, of true statements and terms concerning a primitive concept from which we choose a subset $\sum$ implying T. Choice of $\sum$ should be based on neatness, economy of effort, and the degree of structural insight about T given by $\sum$.

In summary, formal science proceeds as follows. Initially there is a primitive concept which one wishes to investigate. An axiom system is introduced as a model of the concept. Certain key properties of the concept are incorporated as axioms. Initially, the only sense in which the axioms are "true" is that they seem to hold for the concept. The axiom system is studied by purely logical methods to obtain theorems. Conclusions of theorems are only conditional truths; that is, they are true if the axioms are true.

From the point of view of formal science the difficulty with the Euclidean geometry paradigm is almost obvious. Formal science does not study the real world, only conceptualizations of the world. The theorems are not about some phase of the existing world but are about whatever is postulated by thought. Theorems contradicting each other in different axiom systems may merely indicate basic differences in conceptualization. In particular the purpose of the two theories may differ. We have to distinguish between mathematics and its applications.

But there is no mystery, as has sometimes been suggested, about why it is that mathematics is useful for applications. When $\sum$ is motivated by an application then the application is an interpretation of $\sum$ and the theorems of $\sum$ will remain true when interpreted for the application. As Wilder puts it, "What we get by abstraction can be returned."

The subject of applied mathematics consists of proving theorems and constructing counterexamples concerning some primitive concept which we wish to investigate. But the theorems need not be consistent with one another; indeed we will see in our examination of science that the deductive portion of a science is almost always inconsistent. Hence formal science is only a special kind of applied mathematics.

From the previous discussion we observe that knowledge justified exclusively by deduction has two characteristics. First, it is certain, for deduction is truth-preserving. Second, it is conditional, for all theorems are of the form "if A, then B." In particular, that portion of mathematics which is axiomatic method without regard to the validity of the axioms is conditional knowledge. However, much knowledge, particularly scientific knowledge, is unconditional but not certain. Hence, deduction will not, by itself, explain knowledge. Nor will mathematics suffice. Further evidence that some component is missing, is that if $\sum$ is an axiom system and T is the totality of consequences of $\sum$, then, try as we might, we can not prove a consequence outside of T; but scientific knowledge typically grows by observation of new phenomena. A theory which is solely mathematics is not true in any absolute sense, or does it have ethical or moral implications, (such as unbiased, inadmissible, good, best, etc.) which are sometimes attributed to it. The component that is missing is experiment or data. The real-world validity of the whole must be judged by comparing theoretical deduction with actual experiment.

Notes on the Literature

The viewpoint of this chapter is not controversial nor is it hostile to mathematics. Virtually all mathematicians who have examined the foundations of their subject have arrived at this view. The present exposition derives from Frank (1957) and the geometer Blumenthal (1980) but particularly from the point set topologist Wilder (1983).

3

The Evolution of Natural Scientists and Their Theories

3.1. Background

Evidence for the special success of science can be seen everywhere: in the automobiles we drive, the television we view, the increase in life expectancy, and the fact that man has walked on the moon. We may question the values, or lack thereof, of science, but surely science has been singularly successful in many of the enterprises which it has attempted. It is therefore a little surprising that what constitutes meaningful scientific work is one of the most hotly debated issues of our time. The nature of scientific process and method, the aims or goals of science, the nature and objectivity of scientific knowledge—all are controversial. Still, there is much agreement. This chapter discusses the controversy and attempts to find points of agreement.

Most discussions of science and scientific method are implicitly restricted to what might be called natural science, the systemized knowledge of "nature" or the "real world." But there are other concepts of science; for example, mathematics is usually thought to be scientific but much of it is not about nature. The present chapter discusses the controversy in natural science; in Section 13.1 we will have reason to consider more general meanings of the word *science.*

The purely deductive development of natural science from "self-evident truths" was shown to be inadequate by the invention of non-Euclidean geometries. The ingredient which is missing is experience. The pioneer here is Sir Francis Bacon, who emphasized that we learn (acquire knowledge) "by putting nature on the rack," that is, by conducting experiments. Experimentation is asking questions of nature—conditions are imposed and nature's response is observed. In the experimental investigation of natural "laws," events, called conditions, are known or made to occur and outcomes are observed to occur when the experiment is performed, that is when the conditions occur. We may think of an **experiment** abstractly as a doublet (C, Y) where C and Y describe the conditions imposed on nature and the instructions for observation, respectively. An experiment is in fact an experimental situation or set-up, the recipe for performance. A performance or realization of the experiment consists of carrying out the recipe, imposing the conditions, and observing the outcome.

Historically, experiments were first considered to be deterministic in that the outcome can be predicted from the conditions. But all experiments, when examined sufficiently minutely, will be found to be indeterminate: there will be various potential outcomes and the initial conditions will not completely determine the outcome of any particular performance. Chance—to be discussed in Chapters 5 through 7—is a kind of indeterminism where the rules of probability are thought to apply.

We may follow the same recipe but there is no guarantee that we will obtain the same outcome. Nevertheless, it sometimes does happen that repeated performance of an experimental recipe does seem to produce approximately the same outcome and when this does happen we learn an important fact. The most interesting explanation of this is that some condition of the experiment **causes** the observed outcome. Science adopts the Spinoza view, that natural laws exist and that cause of necessity produces effect, but Hume questions the necessity of the cause and effect relationship; Hume insists that law is simply an observed custom in the sequence of events.

Knowledge is not just deduction but neither is it just experience. A modern version of the purely experimental view appears in Persig's *Zen and the Art of Motorcycle Maintenance* (1984, p. 92):

If the cycle goes over a bump and the engine misfires, and then goes over another bump and the engine misfires, and then goes over another bump and the engine misfires, and then goes over a long smooth stretch of road and there is no misfiring, and then goes over a fourth bump and the engine misfires again, one can logically conclude that the misfiring is caused by the bumps. That is induction: reasoning from particular experiences to general truths.

But Persig knows something about motorcycles and roads, and it is with the aid of the theory of the internal combustion engine that he draws his logical conclusion. In the absence of all theory we are equally justified in concluding that the misfiring causes the bumps.

The basis of a purely empirical theory was destroyed by David Hume's demonstration that the principle of induction is a circular argument. Hume's demonstration is as follows. A purely empirical theory of science is based on something close to the **principle of induction**: If conditions C are experienced a large number of times and under a wide variety of other circumstances and each time outcome $Y = y$ is observed then whenever C is experienced, outcome $Y = y$ will be observed. Since this is to be a purely empirical theory, what is the empirical justification for the principle? Presumably the justification is that the principle has worked to establish much of what mankind "knows." It worked to establish the boiling point of water, it worked to establish the relation between the extension of a spring and the weight supported, it works in establishing plants by seed, etc. Therefore the principle will always work. But the justification for this last statement is the principle of induction: The principle has been used to justify itself, a circular argument.

There are other difficulties for the principle; we mention two. First, is the dependence of observation on theory. Usually modern scientific observations are taken with the aid of sophisticated measuring instruments such as optical or electron microscopes or radar; what one is observing through such measurements requires considerable theory to interpret. Second, even after we have decided what we are "seeing" in an observation we must consider the relevancy of other circumstances.

Unfortunately, future experiments (future trials, tomorrow's production) will be affected by environmental conditions (temperature, materials, people) different from those that affect this experiment. It is only by knowledge of the subject matter, possibly aided by further experiments to cover a wider range of conditions that one may decide, with a risk of being wrong, whether the environmental conditions of the future will be near enough the same as those of today to permit the use of results in hand.

Deming (1986, pp. 350–351)

Much of natural science appears to be neither deduction nor experimentation; it seems to be simply accurate description of particulars. But our expectation is that, by describing particulars carefully, general truths will become apparent. "To see what is general in what is particular and what is permanent in what is transitory is the aim of scientific thought," Whitehead (1911, quoted in Feather, 1959, p. 1) aptly observed.

Nature is complex and one cannot rule out the a priori possibility that any detail of a particular is relevant. But the scientist attempts to unravel this complexity by a process of simplification and reasoning checked by experiment. The scientist's simplification of nature is called a theoretical model; the scientist develops it by introducing concepts and their operating rules, which are suggested by common sense and/or experiment, deducing the consequences of such concepts and rules and then checking the whole by comparing with experiment. In this way the scientist sets up an isomorphism between an aspect of nature and the model. Scientific method involves an interplay between deduction and experiment.

As introduction to his dynamical theory of gases, Maxwell (1860) describes the process as follows:

In order to lay the foundation of such investigations on strict mechanical principles, I shall demonstrate the laws of motion of an indefinite number of small, hard, and perfectly elastic spheres acting on one another only during impact. If the properties of such a system of bodies are found to correspond to those of gases, an important physical analogy will be established, which may lead to more accurate knowledge of the properties of matter. If experiments on gases are inconsistent with the hypothesis of these propositions, then our theory though consistent with itself, is proved to be incapable of explaining the phenomena of gases. In either case it is necessary to follow out the consequences of the hypotheses.

Consistency is what mathematics is about. But the analogy may or may not hold for an interpretation. Consider that, if one man can dig a certain hole in 5 days, then 2 men can dig it in 2.5 days. But if one ship can cross the Atlantic ocean in 5 days, can two ships cross it in 2.5 days? The arithmetic analogy holds in the first case but not the second.

Gastwirth (1992) offers this summary: "The purpose of science is not only to describe the world but also to understand the mechanism generating our observations. Thus, scientists develop general theories that explain why certain phenomena are observed and that enable us to predict further results. A theory is corroborated when predictions derived from it are borne out." But Karl Pearson (1935 and 1892, p. 119) insists—

... all these descriptions by mathematical curves in no case represent 'natural laws'. ... They are merely **graduation curves**, mathematical constructs to describe ... what we have observed. ... These scientific formulas **"describe**, they never **explain** ... our perceptions ..."

Pearson's 'graduation curve' is the modern concept of mathematical model.

Franck (1957) summarizes his view as follows: "The goal of science in the twentieth century has been to build up a simple system of principles from which the facts observed by twentieth -century physicists could be mathematically derived." But, from the above examples, we see that the stated goals of natural scientists vary; they construct models to describe, explain, summarize, and/or predict nature. The explanatory power of science satisfies the human need to understand, description records the facts, and summarization makes the whole comprehensible. Prediction facilitates design, arranging things to achieve a desired end.

3.2. "Prediction = Science"?

The title of this section is from Ruhla (1993, p. 1); he continues—

It is often thought that science is an explanation of the world. Though this is an important feature, it is not the most characteristic: **the overriding priority in science is prediction.**

Toulmin (1963) expresses the contrary opinion that natural science is not just "predictivism." The predictivist view is that scientists predict the results of future experiments on the basis of past experiments. Toulmin (1963, p. 115) concludes,

we can never make less than a three-fold demand of science: its explanatory techniques must be not only (in Copernicus' words) 'consistent with the numerical records;' they must also be acceptable-for the time being, at any rate-as 'absolute' and 'pleasing to the mind.'

The title of Toulmin's book, *Foresight and Understanding, An Inquiry into the Aims of Science*," emphasizes that foresight is not enough; understanding or explaining nature is a further task of science. A description by itself, even if it predicts, provides little understanding, Toulmin insists. Examples are a purely empirical regression equation and a multivariate analysis principle component, as in Anderson (1958, Ch. 11), which is an uninterpreted linear combination of the original variables. Predictive descriptions which also explain are the goal.

Toulmin makes a valid point, but we can never be sure of having achieved more than just an analogy and one suspects that part of the difficulty is a breakdown in communication. The word *prediction* as used in scientific theory testing (Table A.7) usually means a deduction or logical consequence and need not imply foresight. The issue in evaluating strength of experimental evidence is not so much the time sequence as the extent to which the experimental outcome suggested the theory. If theory is suggested by experimental outcome then outcome cannot be said to support theory. Historical experiments performed prior to theory formulation are frequently considered to provide strong support for a theory if they are numerous and diverse, for then one has difficulty believing that the theory is ad hoc.

As Toulmin emphasizes, science does strive for understanding, and yet foresight is important too. Experimental evidence which involves foretelling is ideal: if theory precedes experiment then the question of whether the outcome suggested the theory cannot arise. Further, retrospective prediction *is* one kind of explanation. That is, suppose we have an argument that if a theory had been consulted prior to the experiment it would have predicted as likely an event which contains the outcome which did in fact occur. Such an argument is an explanation of the experimental outcome in terms of the theory; and the argument taken together with the outcome constitutes evidence for the theory. Actually, both new and old evidence are not only relevant but essential. According to Feather (1959, pp. 3 and 4), a satisfactory model must—

i.   Summarize in simple form the observed results of a large number of diverse experiments with no exceptions when it should apply;
ii. predict the results for a wider field; and
iii. must not be mere definition (it must be conceivable that the world might have been otherwise); a satisfactory model thus must be falsifiable, to use Popper's terminology.

Ruhla gives a good anecdotal discussion. Human beings have a natural desire to understand the only world in which they live. Hence they seek explanations of why things happen as they do. For example, many of the properties of light may be explained in terms of the theory that light consists of particles. Other circumstances require the theory that light consists of waves. This duality can still be regarded as an explanation if altered circumstances cause light to change state. However, for some interference experiments, as in Ruhla, (1993, Ch. 7), whether light behaves as particles or as waves is determined, according to both theories, at a point before the circumstances are altered. This would imply that the effect precedes the cause, contradicting the causality principle, obeyed by both particles and waves. If, as suggested by Maxwell, scientific knowledge is just an "analogy" then neither the particle analogy nor the wave analogy holds for interference experiments on light and there is no conceptual difficulty. But in fact physicists have regarded these matters as very disconcerting, indicating that they are trying for an explanatory realist theory.

This same investigation provides strong evidence that though they desire and seek explanation and understanding, scientists sometimes have to settle for just prediction. Compared with Einsteinian hidden variable theory, quantum mechanics is a poor explanation but a good predictor. And yet it is now clear that quantum mechanics and the positivist view of Bohr is the clear winner over the realism of Einstein.

A predictive rule p, obtained from a model by logical deduction, will take the form that if initial conditions C are imposed then nature's response will be p(C).

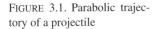

FIGURE 3.1. Parabolic trajectory of a projectile

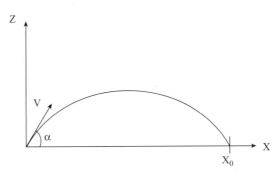

Example 3.1 Ballistics an all too familiar practical example occurs in the science of ballistics, which concerns hitting a target with a projectile. The instructions for performing this experiment will include leveling and orienting the gun; initial conditions will include muzzle velocity v and elevation angle α as in Fig. 3.1. A crude first theory results from the second time derivative assumptions, $z'' = -g$, $x'' = y'' = 0$, where g is the acceleration due to gravity.

Integration yields a parabolic trajectory. Setting $z = 0$, we obtain $y = 0$ and $x = p(\alpha, v) = g^{-1}v^2 \sin 2\alpha$ for predicted coordinates at impact.

Control of nature through adjusting the initial conditions of a predictive model is the cash value of science. If p is a valid predicative rule and we desire to achieve outcome p_0 then we may do so by choosing C to satisfy $p(C) = p_0$. For the ballistics example, if v is known then given a target at range x_0, we may arrange to hit the target by setting $\alpha = 2^{-1}$ arc $\sin(x_0 g/v^2)$.

More importantly, prediction plays a key role in model checking. Agreement of experimental outcome with theoretical prediction is the criteria for the success of a theory. Agreement of theory with experiment entails, first, that the theory must be internally (logically) consistent, and second, that the experiment must be consistent with itself, that is, repeatable. Finally, theory and experiment must agree with each other.

Logical consistency is used here in the technical sense of the calculus of propositions. A proposition is the meaning of a declarative sentence. A collection of propositions is inconsistent if some proposition and its contradiction are both logically deducible from the collection. If a theory implies both p and not p, then no matter which way an experiment turns out it will be inconsistent with the theory.

Nor can theory be consistent with experiment if the experiment is not consistent with itself (repeatable). Predictions must have reference to a process which is in statistical control in the quality control sense. As Shewhart (1931, p. 6) explains—

a phenomenon will be said to be controlled when, through the use of past experience, we can predict, at least within limits, how the phenomenon may be expected to vary in the future. Here it is understood that prediction within limits means that we can state, at least approximately, the probability that the observed phenomenon will fall within the given limits.

The scientific method of theory checking is to compare predictions deduced from a theoretical model with observations on nature. Thus science must predict what happens in nature but it need not explain why. But compare how? Since models aren't right or wrong, only more or less accurate for some purpose, how are we to judge the agreement between model prediction and experimental outcome?

Prior to Kuhn (1962), the checking of hypotheses was thought to be an essentially impersonal matter of logic and experimentation. One simply performed experiments and truth leapt from the test tubes. Occasionally, comparison of theory with experiment **is** direct. Consider a military test officer evaluating the relative merits of two kinds of antiaircraft shell. He fires the two at scrap airplane wings and photographs the results. For the first kind of shell one sees a few isolated holes in the wings; for the second a few struts remain and it is difficult to see that they were once wings. The results leave little room for discussion or need for statistical analysis. More often, whether experiment is consistent with theory is relative to accuracy and purpose. All theories are simplifications of reality and hence no theory will be expected to be a perfect predictor. Theories of statistical inference become relevant to scientific process at precisely this point.

3.3. External Aspects of Science

William James (1842–1910) ventures the pragmatic opinion that " the true is the name of whatever proves itself to be good in the way of belief." On this Will Durant observes, yes, if the proof is that of scientific method, but no, if personal utility is to be the test of truth. Kuhn (1962) seems to agree: "If I am talking at all about intuitions, they are not individual, rather they are the tested and shared possessions of the members of a successful group." Ackoff (1979) relates these ideas to objectivity:

Objectivity is not the absence of value judgments in purposeful behavior. It is the social product of an open interaction of a wide variety of subjective value judgments. Objectivity is a system property of science taken as a whole, not a property of individual researchers or research. It is obtained only when all possible values have been taken into account: hence, like certainty, it is an ideal that science can continually approach but never attain. That which is true works, and it works whatever the values of those who put it to work. It is value-full not value free.

It is now generally agreed that science evolves according to a social mechanism. Evolutionary models of science have been proposed for at least 100 years; for example, in the statistical literature, Box (1980, p. 383) writes, "New knowledge thus evolves by an interplay between dual processes of induction and deduction in which the model is not fixed but is continually developing." Scientists transmit their beliefs to (convince) one another while biological individuals transmit genes.

That science is social is a recent realization. The view of scientific method in this century prior to Kuhn (1962) was that of an automatic process dictated by the internal considerations of logic and experimentation; the role of the scientist was thought to be limited to the creation of new hypotheses. The basis of a science was thought to be a consistent set of axioms which, as a consequence of logical and experimental investigation, were added to in an objective and continuous process that drew ever closer to the truth. The accumulation of scientific facts was often likened to constructing a building by placing one brick of knowledge on another. As Feynman et al. (1975) put it, "The principle of science, the definition, almost, is the following: The test of all knowledge is experiment." Thomas Kuhn (1962) observed that the impersonal asocial view of scientific method is inconsistent with history.

Deduction and experimentation **are** the permissible arguments of scientific testing but the short-term test of scientific truth is peer review. Without loss, we may consider only the example of a professional journal. Matters to be tested are initiated by individual authors; they perform their experiments and submit their prepared arguments to the editors of journals. The primary function of an editor is to organize a good fight and to judge who wins. The disputants (authors, referees, and sometimes the editors themselves) present their arguments in a highly adversarial and biased manner. Hull (1990) documents some of this, with historical evidence from the sciences of evolutionary biology and of systematics (the science of classifying living things). The editor or perhaps an editorial board decides whether the submission will be published. Thus, short-term truth is whatever emerges from a staged battle. An idea becomes long-term scientific knowledge if (i) the author of the idea wins the publication battle and (ii) the idea proves resistant to refutation by the peer review process. Scientists must put their professional reputations at risk and argue aggressively in order to get their ideas accepted.

Kuhn observed that important events in science just didn't happen according to the impersonal continuous growth theory. He claimed that science consists instead of relatively calm periods of agreement on world view alternating with periods of "revolution" during which the old view is replaced by an incompatible new one. The theories of a scientific discipline are typically not developed from a single set of axioms; rather the hypotheses of the derivations are altered as necessary to obtain results. As a consequence, sciences are typically inconsistent at all times, but scientists will constantly be striving to eliminate as much of this inconsistency as possible. Kuhn concludes that science is not an impersonal and continuous accumulation of facts drawing ever closer to the truth, but rather, when viewed from afar, a community and period phenomenon much like art.

Concepts central to Kuhn's now familiar scheme of things are paradigm, normal science, and anomaly. A paradigm may be initially described as a conceptual world-view. During periods of normal science the paradigm is not being questioned but is being extended to a new situation or problem. Normal science does not aim at novelties of fact or theory and when successful, finds none. Nevertheless, novelties are repeatedly uncovered by the activities of normal science. The way that this comes about is that anomalies, violations of expectation, arise in the research of normal science. A crisis in normal science, caused by repeated and insistent anomaly, leads to periods of revolution in which the old paradigm is discarded and new theories are invented. (Toulmin [1963] agrees that the anomalies are most in need of explaining.) Kuhn's picture of the advance of science follows the scheme: prescience—normal science—crisis—revolution—new normal science—new crisis.

Since 1962, those who would understand science have emphasized a greater role for external factors. They have increasingly resorted to social, economic, political, and subjective explanations. An extreme view is that facts aren't discovered, they are negotiated.

Toulmin (1972) also emphasizes that science is a social process that cannot be explained in terms of a purely internal scientific method. One role of the scientist in science is that of innovator, contributor of original ideas, but this originality is everywhere constrained within a particular conceptual inheritance. The articulation of individual thoughts presupposes the existence of a shared language and is possible only within the framework of shared concepts. But the scientist's second role of tester, carrier, teacher, and test case for ideas is also important for the acceptance and propagation of scientific ideas.

The primary effect of Kuhn's book has been to create a Kuhnian crisis, which still exists, for scientific method. Critics have objected, first, that Kuhn's description does not really provide a theory of how science works, and second, that it over emphasizes the role of the decisions and choices of scientists. It makes of physics the study of the psychology and sociology of physicists rather than the study of matter. No doubt, science is a social practice, but surely its findings are largely about nature.

The issue of over emphasis on decisions has risen in statistics as well. Tukey (1960) calls attention to a distinction between decisions and conclusions—a distinction which he sees as important for understanding statistical inference. He locates his discussion in the realm of science; while scientists make many decisions, a scientific body of knowledge accumulates primarily by arriving at conclusions. The decisions of decision theory are of the form "let us decide to act for the present as if." Conclusions on the other hand—

are established with careful regard to evidence, but without regard to consequences of specific actions in specific circumstances. . . . Conclusions are withheld until adequate evidence has accumulated.

A conclusion is a statement which is to be accepted as applicable to the conditions of an experiment or observation unless and until unusually strong evidence to the contrary arises. This definition has three crucial parts, . . . It emphasizes "acceptance", in the original, strong sense of that word; it speaks of "unusually strong evidence"; and it implies the possibility of later rejection. . . .

First the conclusion is to be accepted. It is taken into the body of knowledge, . . . Indeed, the conclusion is to remain accepted, unless and until unusually strong evidence to the contrary arises. . . . Third, a conclusion is accepted subject to future rejection, when and if the evidence against it becomes strong enough. . . . It has been wisely said that "science is the use of alternative working hypotheses." Wise scientists use great care and skill in selecting the bundle of alternative working hypotheses they use.

Tukey (1960, pp. 425–6)

The statistical community appears to have accepted Tukey's distinction between decisions and conclusions but has not followed up by developing his conclusion theory; that task takes the statistician out of his comfortable area of expertise.

Note that, in agreement with Toulmin's quote of Copernicus (Section 3.2), Tukey conclusions are either in or out of the bundle of working hypotheses, and the only possible actions are adding or dropping conclusions. Scientific conclusions cannot be accepted with certainty, but when they are accepted, it is wholehearted and without reservation—for the time being. The reason for this is that deduction— and particularly mathematics—plays an important role in science; and classical logic, the tool used for deduction, adopts the simplification that statements are either true or false (in or out of the bundle of working hypotheses). Partially true statements cannot be treated by classical logic and therefore do not mesh with the deductive aspect of science.

Richards (1987, Appendix I) presents an account of models—(1)static,(2) growth, (3) revolutionary, (4) Gestalt, and (5) social-psychological—which have historically been important "concerning the character of science, its advance and the nature of scientific knowing." He then turns to evolutionary models concluding with his own version. Richards compares evolutionary models favorably with the above five and with Lakatos's scientific research programs.

Giere (1992) presents a decision-theoretic model for science which allows an evolutionary view, but he says he is not there concerned to develop the evolutionary analogy.

Toulmin (1972) argues that cultural and biological evolution are special cases of a more general selection process. Richards (1987, p. 578) criticizes Toulmin for abandoning the specific "device of natural selection so quickly," but what is the special virtue of an analogy with natural selection? We think that any satisfactory evolutionary treatment of science must incorporate Toulmin's feature since a major objection to the evolutionary comparison is that science is, to some extent, intentional, whereas natural selection is not. Giere (1992) says that science has a substantial cognitive component that does not fit well with the evolutionary model. But artificial breeding has a cognitive component and is highly intentional, and artificial and natural selection manifest themselves according to the same pattern. A major part of Darwin's (1859) argument for natural selection is that features can be bred into plants and animals by artificial selection; indeed the very name "natural selection" was coined to make this analogy. (I, thus, will argue that the proper biological analog to the selection of ideas is the selection of genes by both natural and artificial means.)

Building on Toulmin, Hull (1990) makes a convincing case that science must proceed according to a process something like biological evolution. Hull's book contains a unified analysis of selection processes as suggested by Toulmin. Hull is interested in the sociology of science. The behavior of scientists may be explained in terms of curiosity and a desire for recognition. "Science is so organized that self interest promotes the greater good." But he sees no reason why everyone need do epistemology all the time, and he suggests that no justification of scientific knowledge or method exists in terms of internal principles.

Hull argues that the tension between cooperation and competition is the mechanism that makes science go and that quite a bit of this mechanism is to be explained in terms of curiosity, credit (attribution), and checking. The explanatory power of science satisfies curiosity. The desire for credit (preferably with a formal acknowledgement) is half of the remaining dynamic. The mutual checking of research that goes on in science is the other half. Hull further explains (p. 319) that

The most important sort of cooperation that occurs in science is the use of the results of other scientists' research. This use is the most important sort of credit that one scientist can give another. Scientists want their work to be acknowledged as original, but for that it must be acknowledged. Their views must be accepted. For such acceptance, they need the support of other scientists. One way to gain this support is to show that one's own work rests solidly on preceding research. The desire for credit and the need for support frequently come into conflict. One cannot gain support from particular work unless one cites it, and this citation automatically confers worth on the work cited and detracts from one's own originality.

The other side of desiring credit is the desire to avoid blame and the need to check one's speculations. But as Hull remarks, the self correction which is so important to science does not depend solely on individual unbiasedness or objectivity: "Scientists rarely refute their own pet hypotheses, especially after they have appeared in print, but that is all right. Their fellow scientists will be happy to expose these hypotheses to severe testing."

Because of the logical situation of Table A.7, that the falsity but not the truth of generalizations can be deduced from appropriate particulars, Popper suggests that science learns primarily from its mistakes through the falsification of bold conjectures. But this view is challenged by Chalmers (1982, p. 55): "If a bold conjecture is falsified, then all that is learnt is that yet another crazy idea has been proven wrong."

Chalmers suggests instead that the most convincing form of theory—experiment consistency—is a successful novel prediction.

A confirmation will confer some high degree of merit on a theory if that confirmation resulted from the testing of a novel prediction. That is, confirmation will be significant if it is established that it is unlikely to eventuate in the light of the background knowledge of the time.

Chalmers (1982, p. 58)

A prediction of theory T will be called conservative or novel according as it is probable or improbable on the basis of background knowledge, excluding T. Evidence for a theory consists of a successful prediction which follows readily from the theory, but is difficult to explain if the theory were not true.

Chalmers' view, called modern falsificationism, is supported by the opinion of Kuhn (1962) that the single most effective claim, in bringing about convergence to a new paradigm, is probably that the new paradigm can solve the problems that led the old one to crisis. A successful novel prediction not only creates an anomaly for background knowledge but also suggests how the anomaly might be resolved.

We have been emphasizing the controversy in natural science but perhaps we should end on a more unifying note; some broad characterizations are generally agreed on. Scientific method is an approach to knowing, which emphasizes system and objectivity. The system involves observing, wondering, and thinking about things. Most observations result from experiments, which are questions put to nature. The most common experimental setup is the comparison, for example a clinical trial of a cold remedy versus a placebo. Scientific thinking is directed toward finding patterns in our observations and explaining what we observed. The former process of inference is called induction. Scientific thinking takes the forms of imagining or theorizing and constructing simplifying models.

Science is cyclical and self-perpetuating since observations will not always agree with theory. This creates a problem of misfit which leads to further wondering and thinking which motivates further observations to check the results of new thought, etc. In this sense science is problem solving. The objectivity of science results not from the high moral professional behavior of scientists but from the resistance of logic and nature.

3.4. Conclusion Theory

Kuhn (1962), Toulmin (1972), Richards (1987), and Hull (1990) document and analyze many historical instances of science. Sections 3.1–3.3 reviewed and discussed their and other diverse views. But if we are to adapt statistics to aid science then a single model of science is needed. This section explains the attitude toward natural science to which we will attempt to conform.

A natural science is a bundle of trustworthy beliefs about nature; they are trustworthy by virtue of having been judged to agree, in the past, with experiment. The beliefs of science are Tukey conclusions; they have the status of working hypotheses. A **conclusion** is a statement which is to be accepted as applicable to the conditions of an experiment.

Science evolves according to a social mechanism. Scientists convince (transmit their conclusions to) one another, just as biological individuals transmit genes. Regularities do exist in nature and conclusions can be checked; some conclusions are just better able to withstand experimental test and subsequent review than others. Scientists who hold these more robust views are more successful in solving scientific problems, achieving publication and winning support. Theories evolve in the manner in which genetic traits evolve: primarily through the professional survival and publication of the scientists who carry them.

A "population" of scientists is a collection of individuals grouped together over time because they function as a unit with regard to transmission of some class of conclusions. The essential characteristic of a population is closure: the conclusions of an individual of a population will have been transmitted from earlier individual(s) of the same population. The subset of individuals of a population that exist at a specific time may be called a "generation" of the population. A population having a high degree of similarity with respect to conclusions is a scientific school. The collective conclusions of a school are a tradition. When communication between two populations of scientists breaks down to the point where, as Kuhn says, they are "practicing their craft in different worlds" and transmission of conclusion from one to the other no longer occurs, then they are two different schools.

Perhaps the major conceptual hurdle in thinking about the evolution of science is that the individual scientists must be considered in their scientific capacity. Scientists are organisms and survival of the organism is necessary for the survival of the scientist, but professional survival is what is relevant to the scientific process. Hull (1990, p. 434) writes, "The ideas that these scientists hold do not produce them in the way that genes produce organisms. . . " But although ideas do not produce scientists as organisms, ideas do indeed produce scientists as scientists.

The professional environment in which a scientist works has several aspects. There is the specific location, the type of employer (whether university, industry, or perhaps government), the social and political climate, etc., but by environment, we mean also the kind of natural phenomena to be described, explained, summarized, and/or predicted—at the coarsest level, the discipline, whether it is psychology or chemistry, and within the subject, the specialty. Here, environment is interpreted primarily as a class of problems that the scientist attempts to solve.

Basic assumptions are—

i.   Variation of conclusion: different scientists in a population will have accepted different conclusions.
ii. Differential fitness: scientists holding different conclusions—different bundles of working hypotheses—will have different rates of survival (as scientists) and different rates of success in transmitting (convincing, publishing) their conclusions to other scientists in different environments.

Hull is correct about credit received being important to scientific process. If a scientist does not receive some credit, then he or she will be unable to survive as a scientist. "On the view of science," Hull writes (p. 361), "that I am developing, success and failure is a function of the transmission of one's views, preferably accompanied by an explicit acknowledgement."

Fitness of conclusion for an environment is intrinsic power of ideas to describe, summarize, explain, and/or predict the results of experiments in that environment. It is in this way that the internal aspects, discussed in Sections 3.1 and 3.2, are relevant. The reason that fit conclusion has reproductive potential is that these same goals are the criteria for success of a theory. Individuals with fit conclusions will solve new problems, thus earning funding, publications, positions of authority, or at least the respect of their population; their superior explanations and demonstrated success tend to convince their colleagues.

The collective conclusions of a population of scientists are called its "belief pool." The belief pool will not extend over all conclusions, only those relevant to a class of experiments or problems. As A.J. Ayer (1982, p. 11) puts it, "the success of scientific experiments depends on our being able to treat small numbers of our beliefs as isolated from the rest."

According to our basic assumptions some individuals will have a higher tendency to survive *as scientists* and transmit their particular conclusions to subsequent generations. Hence, as long as the class of problems considered does not change, succeeding generations will likely contain a higher proportion of the more fit conclusion configurations. In the course of time the belief pool of the population will become dominated by the conclusion structure of the fittest configurations, thus, perhaps, replacing the traditionalists by a new school. Over longer time, since most scientists succeed by specialization, the total scientific environment will be partitioned among specialties, each particularly adapted to some portion. This explains the proliferation of scientific schools. Developed schools are Kuhn's normal sciences.

Individual scientists are the carriers of conclusions and theories, but conclusions have an existence of their own, which is continuous with all scientists past and present. There is more to conclusion than just individual conclusion—there is articulated conclusion, shared conclusion, and stored conclusion, but most important in the present context, is transmitted conclusion. Minds die, but conclusions continue until they disappear from the belief pool.

Nevertheless conclusions require minds to transmit them, and minds contribute the markedly original ideas which we shall call mutations of conclusion. Most mutations are quickly found lacking and enter the belief pool only briefly, but occasionally a mutation with latent fitness, called an exemplar, comes along. Exemplars demonstrate their superiority, survive, and in the course of time their implications develop into theories. In the sense that a chicken is an egg's way of producing a new egg, theories beget new ideas, a few of which become exemplars of new theories that may replace their "parents." It is exemplars that are primarily responsible for major adaptive theory change.

At least two kinds of conclusions are relevant to science—personal conclusions and the social conclusions of a population. A scientist's conclusion is the state of mind that a statement *is* applicable to an experiment. A scientific conclusion is an agreement—of sorts—among scientists. The test of scientific truth is not raw experiment but peer review—a staged confrontation of author(s) versus referees and editors. Experiments do not speak for themselves; they have to be interpreted in the context of some theory and the interpretation promoted and defended. An idea becomes scientific knowledge if (1) the author wins the publication battle and (2) the idea proves resistant to refutation by the peer review process. The current conclusions of a science are largely contained in its unrefuted journal articles.

Personal conclusions are unimportant to science but crucial for scientific process since they become the social conclusions which are the content of a science. It is not the personal considerations of what to believe or how to act which are central to scientific process but rather the social matter of transmission of conclusion from one scientist to another. The central question of scientific method is: How does one scientist convince another of his conclusions? A scientist cannot force his or her colleagues to accept a conclusion; he can only confront them with evidence—reasons to believe.

We close with a listing of the desirable characteristics of conclusion theory; the first three are shared with most evolutionary models.

a. preserve "the traditional distinction between the process of discovery, when ideas are generated, and generational criteria are continually adjusted, and the process of justification, when ideas are selected" (Richards, 1987, p. 576). Individuals generate mutations, selection is by peer review.

b. "avoid the presumption that theories are demonstrated by experience" and "that theories and creative ideas arise from any sort of logical induction from observations; thus, the older and newer problems of induction are skirted" (Richards, p. 576). Conclusions are just hypotheses to be worked with unless and until unusually strong evidence to the contrary arises.

c. restrain the destructive relativism of the social-psycho-logical model while preserving the edge of its insight (Richards, p. 574). The test of truth is peer review but logic and experiment are the possible arguments.

d. allow a needed mix of what we have called the internal and external aspects of science.

e. explain why Kuhn's historical observations do not make science subjective: the individual scientist's decision to accept a conclusion is indeed subjective and will be based, in part, on personal utility, but whether that conclusion survives to propagate (through its adherents)—its long-run public acceptance—depends on its fitness.

f. preserve the objectivity of science: all scientific conclusions are not equally valid. Regularities do exist in nature, and conclusions can be checked; some just turn out to be more fit than others.

g. distinguish science from nonscience. The criteria are the old hurdles of logical consistency and agreement with experimental outcome.

h. explain the errors of science as well as its special success: scientific truth is only trustworthy conclusion, but science has been restricted to those aspects of human investigation for which there is a relatively impartial way of checking.

4

Law and Learning

Like mathematics, statistics, too, is a handmaiden of science. See the preface in Senn (2003) in this regard. But, perhaps unlike mathematics, statistics derives much of its character and meaning from related disciplines. In addition to logic, mathematics, and scientific method, there are other bodies of thought to which statistics can look for guidance and with which it need come to terms. Further important guides are law, learning theory, and economics. Here we mention the first two; later, in discussing the price interpretation of probability, we touch on economics.

4.1. The Law

Since ancient times the law has struggled with concepts of evidence, proof, fact-finding, and decision. We do not try to summarize how the law works; that would be beyond our capacity and too great a digression. We do make the point that, while legal truth-finding has points in common with the scientific peer review process, law relies less on logic than does science. Justice Fortas, in Irons and Guitton (1993, p. 187) describes the legal approach:

... our adversary system, ... means that counsel for the state will do his best within the limits of fairness and honor and decency to present the case for the state, and counsel for the defense will do his best, similarly, to present the best case possible for the defendant, and from that clash will emerge the truth.

That law and science differ is clear from

the words of Justice Oliver Wendell Holmes, who noted a century ago that "the life of the law has not been logic; it has been experience." Holmes also reminded us that "the prejudices which judges share with their fellow-men" have had a great influence "in determining the rules by which men should be governed."

Irons and Guitton (1993, p. 5)

and further from Donald Sullivan:

Typically (defendants will) assert a variety of different and often contradictory things. And under our law it's permissible to do that.... It's kind of like the situation where somebody says, I'm suing you because your dog bit me, and the guy will come in and defend and say, my dog didn't bite you, besides, my dog is real friendly and never bites anybody, and defense number three, I don't have a dog.

Irons and Guitton (1993, p. xvii)

Some legal views, if accepted, have direct implications for statistics. More often the connection is indirect and implications difficult to see. Reasonable people can and do look at these issues in different ways. Our main source for the legal perspective is Eggleston (1978). Sir Richard Eggleston, Chancellor of Monash University, writes from the perspective of the English legal scholar; but he implies a multinational validity. His book appears in the series *Law in Context*.

Eggleston states that probable has the precise legal meaning of having probability greater than one-half.

The law treats the issue of what constitutes valid evidence in a seemingly round about way. Evidence is "relevant" if it makes a fact at issue more or less probable. This just says that the conditional probability of the fact calculated assuming the evidence does not equal its unconditional probability or that fact and evidence are not probabilistically independent. However, some evidence which is relevant is "inadmissible," that is, will not be listened to by the court. Hearsay evidence is one example—it violates the principle that only the best evidence bearing on a point is admissible; the testimony of a first-hand observer would presumably be more reliable. A second example is evidence concerning character. In particular, evidence of a prior conviction increases probability of guilt and hence is relevant but is not admissible. The general principle here is that evidence which is relevant merely by reason of general similarity is inadmissible. An independent basis for postulating a causal relation between the evidence and the fact is needed.

The importance of fact-finding and credibility is that

In most litigation, the decision depends on which of two contradictory versions of the facts should be believed.

Eggleston (1978, p. 137)

Thus, significance testing—to be discussed in Chapters 10 and 11—should prove useful. Judges take the following factors into account when judging truthfulness: (i) compatibility, (ii) indications of general reliability or unreliability of the witness, and (iii) inherent probability. By compatibility is meant *internal* consistency of the testimony and compatibility with other witnesses and with undisputed facts. Unlike science, consistency of observation with hypothesis—for example, whether symptoms are more consistent with heart failure or asphyxiation—is not to be a criterion for judging truthfulness.

Eggleston (p. 89) describes a double standard concerning burden of proof. "The general rule in civil cases is that the burden of proving that a fact exists rests on the person who asserts it." Civil cases are decided on a more probable than not basis. "In criminal cases, the burden is on the prosecution to prove the guilt of the accused 'beyond reasonable doubt.' "

The legal view of fact-finding and prediction is interesting:

... from the point of view of the evaluation of probabilities, past but unknown facts are in the same position as future facts. Nevertheless, the law draws a sharp distinction between the evaluation of probabilities in relation to past facts, and prediction as to what is likely to happen in the future.

In relation to past facts, when a judge finds that the necessary standard of proof has been achieved ... he is entitled to treat the facts as established, and to give a decision on the footing that those facts exist. The decision will be the same whether the facts are established with absolute certainty or on the balance of probability.

When, however, the question is how much the plaintiff should be paid for prospective loss in the future, the law takes a different view. If there is a doubt as to whether a particular event will occur in the future, and the amount of the plaintiff's damages will be affected by the happening of the event, the court will make allowance for any measure of uncertainty in the prediction. Thus, where there is a forty per cent probability that the plaintiff will have to undergo an operation at a future time, the plaintiff will receive the 'present value' of forty per cent of the estimated cost of the operation.

Eggleston (1978, p. 72)

Thus, legal fact-finding seems to assume a two-valued or accept–reject logic, whereas prediction calls for a mathematical expectation. Eggleston provides us with a qualifying observation.

Since judges never really know the truth, but only arrive at conclusions as to what the truth would turn out to be if only they could know it, the fact-finding process has much more affinity with prediction than judges are prepared to recognize. What the judge is required to do is to make a prediction as to how the facts would turn out if he were vouchsafed divine guidance into the truth, and in making such a prediction he is entitled to rely on the probabilities.

Eggleston (1978, p. 26)

4.2. Learning Theory

The scientific method is a special way of learning about the world through experience and many of us have spent years learning the facts of science. Hence it is tempting to use our own personal educational experiences as a guide to scientific process. We therefore briefly consider science in the context of learning.

Psychological learning theory is about individual learning. There is no single view of how learning takes place, but some features generally are accepted by psychologists (see Hergenhahn, 1988). A popular definition of learning, suggested by Kimble (1961, p. 6), is that it is a relatively permanent change in behavioral potentiality that occurs as a result of reinforced practice. An expectancy that is confirmed consistently develops into what commonly is called a belief; but belief cannot be observed, so learning theory emphasizes behavior.

Most learning is by imitation or association. Mother seats the child in a wooden support and says the word *chair*. Through repetition the child comes to associate the word *chair* with the wooden object. Much of the learning of a science, like the above illustration, consists of learning to speak the language; examples are anatomical parts of the body and names of chemicals.

Other scientific learning is gained through insight rather than reinforced practice. As Hergenhahn (1988, p. 257) puts it—

Insightful learning usually is regarded as having four characteristics:

1. the transition from presolution to solution is sudden and complete;
2. performance based upon a solution gained by insight usually is smooth and free of errors;
3. a solution to a problem gained by insight is retained for a considerable length of time; and
4. a principle gained by insight is applied easily to other problems.

Particularly relevant to science is a psychological theory of learning dev eloped by E.C. Tolman (1886–1959). Hergenhahn (1988, p. 296) summarizes Tol man's theory as follows:

1. The organism brings to a problem solving situation various hypotheses that he may utilize in attempting to solve the problem.
2. The hypotheses which survive are those that correspond best with reality; that is, those that result in goal achievement.
3. After a while a clearly established cognitive map develops that can be used under altered conditions. For instance, when his preferred path is blocked, the organism simply chooses, in accordance with the principle of least effort, an alternative path from his cognitive map.
4. When there is some sort of demand or motive to be satisfied, the organism will make use of the information in its cognitive map.

The fact that information can exist but only be utilized on demand is the reason for the word *potentiality* in Kimble's definition. The cognitive map is Tolman's explanation of why a principle gained by insight is applied easily to other problems. For science, the cognitive map often is deductive or mathematical, and reinforced practice consists of experimentation that is arranged rather than encountered.

Most of what passes for individual scientific knowledge, whether it is language or problem-solving, is based on imitation. Perhaps the usual method of scientific teaching is by example. The teacher formulates and solves a problem and then assigns a similar problem that the student is to structure and approach in an analogous manner. The other side of teaching by example is learning by imitation. Very little of what a scientist "knows" is the result of personal experience. Evidence is grounds for belief. The most commonly used reason for believing a statement is expert judgment—someone we trust states the truth of the statement. Citation of an authority is the most common form of evidence.

It is sometimes remarked that a student can gain insight into the scientific process by taking a laboratory science, but there too the learning is by imitation. Any discrepancy (and there is always some discrepancy) between theory and experimental outcome reflects on the student's poor experimental technique rather than the truth of the theory. Only at a much later graduate research level will a discrepancy be given serious scientific consideration.

Hence, much individual scientific knowledge, since it is based on imitation, is tradition, which is culture dependent; the individual has no personal basis for judging its truth. Education in science, the learning which goes on when an individual studies a scientific tradition, is not then the process by which mankind learns to predict a new phenomena. Neither scientific knowledge nor process is personal. Presumably interested students can retrace the evidence for a belief and decide its truth for themselves, but typically this is practical only at the graduate research level and then only a few beliefs can be scrutinized. This situation raises interesting questions (discussed elsewhere in this and the previous Chapter) of whether, why and in what sense a scientific tradition is true.

Tradition is nevertheless valuable, it permits one individual to benefit from the experience of another and for society as a whole to transmit a useful trait to a later generation. The success of a human individual tends to be determined by the traditions he inherits, the culture to which she belongs. The success of mankind can be attributed to the extent to which the possibilities of learning through speech have been exploited. Speech, the representation of things and the relations between things in verbal symbols, provides an efficient means of analyzing and transmitting experience. Tradition based on speech facilitates flexible and rapid problem solving and permits better predictions of effects from causes. The possibilities opened by the technique of a spoken language are limited only by the scope of the language and the capacity of the brain to store and process verbal symbols. This has led on the one hand, to the invention of mathematical and logical languages with more specialized scope and, on the other, to the invention of writing, printing and computers in order to extend the role of tradition beyond individual human memory and capacity. But in the end, we must carefully distinguish scientific process from science education.

II

Interpreting Probability

5

Introduction to Probability

Gambling is the origin of the mathematical theory of probability. Cardano, who died in 1576, wrote a 15-page "gambler's manual" which treated dice and other problems; however, the effective beginning of the subject was in the year 1654. The Chevalier de Mere was concerned over the following problem of "points": A game between two persons is won by the player who first scores three points. Each of the participants places at stake 32 pistoles, the winner taking the entire stake of 64 pistoles. If the two leave off playing when the game is only partly finished, then how should the stakes be divided? For example, if the players have one and two points, respectively, and their chances for winning each point are equal, then what should be the division of stakes? The Chevalier consulted Blaise Pascal, who solved the problem and communicated his solution to Fermat. In the ensuing correspondence, the two mathematicians initiated the study of probability theory. In Chapter 6 we shall return to the modern theory of gambling.

We begin by discussing the familiar ideas of elementary probability, but with different emphasis. At an advanced level probability is called a measure; but *what* does probability measure? That will be the concern here. After reviewing aspects of the modern axiomatic theory—to establish notation and terminology—we turn to motivation and interpretation of the axioms. Earlier concepts, such as "equally likely" and frequency probability guided the choice of axioms. The earlier concepts are still important because they provide interpretations of abstract mathematical probability, just as light rays and stretched strings are physical realizations of the geometric straight line. In Chapter 2 we saw that geometry played a key role in understanding what mathematics is about. In geometry one starts with undefined concepts such as point and line, makes assumptions (called axioms) about these concepts, and uses the assumptions to prove theorems. An important strength of mathematical disciplines is their capacity to describe many seemingly unrelated aspects of nature. Axiomatic probability is pure mathematics; hence, it will not be surprising to find that the same axioms will be subject to several interpretations.

In this chapter we will introduce the following concepts of probability: (i) the probability of event A is #A/#S, where #A is the number of elementary events in A and #S is the total number of elementary events; (ii) probability is long-term relative frequency; and (iii) probability is a formal science aimed at measuring "tendency to happen." Then in Chapter 6 probability is derived as perceived fair price. Section 6.2 treats probability as a measure of personal degree of belief.

5.1. The Axiomatic Theory

The undefined concepts of probability theory are events and probability itself. Let S, the universal set, be a collection of elementary events and α a class of subsets of S. In statistics, where one is concerned with drawing inferences from samples, it is customary to call an elementary event a **sample point** s. The collection of all sample points $S = \{s\}$ is then called the **sample space**.

Kolmogorov (1950) is credited with providing the axiomatic basis of probability; his axioms—in the finite addition case—are as follows:

I. α is closed[1] under the operations of union, intersection and complementation. That is, if A and B are in α then $A \cup B$, $A \cap B$ and $\bar{A}$, the complement of A, are also in α.

II.   S is in α.

III. To each set A in α is assigned a non-negative real number P(A), called the probability of A.

IV. $P(S) = 1$.

V.   (Finite addition axiom) If A and B have no element in common,[2] then $P(A \cup B) = P(A) + P(B)$.

A system of sets α together with a definite assignment of numbers P(A), satisfying the above axioms is called a probability P.

Several consequences of these axioms are immediately apparent. First, if $A_1, \ldots, A_n$ are finitely many mutually exclusive events then $P(A_1 \cup \ldots \cup A_n) = P(A_1) + \cdots + P(A_n)$. Second, for any event C, from $C \cup \bar{C} = S$ and Axioms IV and V, it follows that $P(\bar{C}) = 1 - P(C)$ and in particular $P(\Phi) = 0$. Also $P(C) = 1 - P(\bar{C}) \leq 1$. The general formula for the probability of a union is $P(A \cup B) = P(A) + P(B) - P(A \cap B)$. This follows from V as a consequence of $P(A \cup B) = P(A) + P(B \cap \bar{A})$ and $P(B) = P(B \cap \bar{A}) + P(B \cap A)$.

If $P(C) > 0$, then Kolmogorov calls the ratio $P(B|C) = P(B \cap C)/P(C)$ the **conditional probability** of B given C. Alternatively, we may introduce conditional probability through the **multiplication rule**:

$$P(B \cap C) = P(B|C)P(C).$$

The multiplication rule generalizes, by mathematical induction, to

$$P(A_1 \cap A_2 \cap \ldots \cap A_n) = P(A_1)P(A_2|A_1)\ldots.P(A_n|A_1 \cap \ldots \cap A_{n-1}).$$

---

[1] A class is closed under an operation if performing the operation on things in the class does not yield things outside the class.

[2] If $A \cap B = \Phi$, that is A and B have no element in common, then A and B are called disjoint sets or mutually exclusive events.

For C a fixed set with $P(C) > 0$, we easily see that $\alpha C = \{A \cap C : A \in \alpha\}$ is a closed class of subsets of C, $C \in \alpha C$, $P(B|C) \geq 0$, $P(C|C) = 1$, $P(A \cup B|C) = P(A|C) + P(B|C)$ provided $A \cap B \cap C = \Phi$, $P(A \cap B|C) = P(A|B \cap C)P(B|C)$ and $P(A \cap C|C) = P(A|C)$. Hence conditional probability is probability in the earlier sense with reduced sample space C and reduced class of subsets α C. On the other hand, from the multiplication rule we may write $P(A) = P(A \cap S) = P(A|S)P(S) = P(A|S)$, and hence all Kolmogorov probabilities may be viewed as conditional.

Events B and C are called independent, if $P(B \cap C) = P(B) \cdot P(C)$. Three events A, B, and C are said to be independent, if they are independent in pairs and also $P(A \cap B \cap C) = P(A) \cdot P(B) \cdot P(C)$. Similarly n events are **independent**, if for each choice of n or fewer of the events, individual probabilities multiply to yield the probability of the intersection.

The finite addition axiom, V, will suffice for many elementary cases but soon one is confronted with unions of infinitely many events and a satisfactory theory requires a countable additivity axiom (Axiom V′): If $A_1, A_2, \ldots$ are pairwise mutually exclusive events (finite or denumerable in number), then

$$P(A_1 \cup A_2 \cup \ldots) = P(A_1) + P(A_2) + \cdots.$$

A set function $P(C)$, defined for a class α of sets closed under countable set operations, and satisfying Axioms I–IV and V′ is called a **probability measure**.

The motivation for requiring that α be closed under countable set operations is that we wish to discuss events having structure of this complexity; for example, $A = A_1 \cup A_2 \cup \ldots$ and $B = \cap_{i=1}^{\infty} B_i$ where A_j and B_i are events. The closure of α then insures that $P(A)$ and $P(B)$ are defined.

The importance of the countable additivity of probabilities may be seen from its equivalence to the continuity property. A probability measure is said to have the **continuity property** if

$$\lim_{k \to \infty} P(B_k) = P\left(\bigcap_{k=1}^{\infty} B_k\right),$$

whenever $\{B_k\}$ is a nested decreasing sequence of events.

Theorem 5.1 In the presence of the finite addition axioms the continuity property is equivalent to Axiom V′.

Proof

i. Axiom V′ implies the continuity property. Let $B_1 \supset B_2 \supset \ldots \supset B$, where $B = \cap_{k=1}^{\infty} B_k$, then

$$B_k - B = (B_k - B_{k+1}) \cup (B_{k+1} - B_{k+2}) \cup \ldots,$$

and from Axiom V′

$$P(B_k) - P(B) = P(B_k - B) = P(B_k - B_{k+1}) + P(B_{k+1} - B_{k+2}) + \cdots.$$

In particular, $P(B_1) - P(B) = P(B_1 - B_2) + P(B_2 - B_3) + \cdots$, and the series on the right converges so that its remainder must approach 0; but the remainder after $k - 1$ terms is $P(B_k) - P(B)$. Thus $\lim_{k\to\infty} P(B_k) = P(B)$.

ii. Conversely, the continuity Property implies Axiom V'. For, if $A_1, A_2, \ldots$ are pair wise mutually exclusive events then define $B_k = \bigcup_{i=k}^{\infty} A_i$. Now, from finite additivity,

$$P(A_1 \cup A_2 \cup \ldots) = P(B_k) + P(A_1) + \cdots + P(A_{k-1})$$
$$= \lim_{k\to\infty} P(B_k) + P(A_1) + (A_2) + \cdots .$$

But $\bigcap_{k=1}^{\infty} B_k = \Phi$, for if $s \in \bigcap_{k=1}^{\infty} B_k$, then s is an element of one of the sets $\{A_i\}$, say $s \in A_n$. Then,

$$s \notin A_{n+k} \text{ for } k = 1, 2, \ldots,$$

because A_n and A_{n+k} are disjoint. Now $s \notin B_{n+1}$ and $s \in \bigcap_{k=1}^{\infty} B_k$, which is a contradiction. The continuity property allows us to conclude the proof as follows:

$$\lim_{k\to\infty} P(B_k) = P\left(\bigcap_{k=1}^{\infty} B_k\right) = P(\Phi) = 0.$$

A real-valued **random variable** $X(\cdot)$ is a mapping from the sample space to the real line. An example is the indicator function of an event A, defined for each sample point s in the sample space as follows:

$$I_A(s) = \begin{cases} 1, s \in A \\ 0, s \notin A \end{cases}.$$

As is common in probability theory, when discussing random variables we often suppress the role of sample point and space, writing X instead of $X(\cdot)$.

The **distribution function** $F(x)$ of a real random variable X is, by definition, $F(x) = P(X \leq x)$. An alternative way of describing the probabilistic behavior of a random variable is to give the **density** function. Here the elementary theory divides into two parts according as X (or its distribution) is discrete or continuous. X is **discrete** if it can assume at most a denumerable number of values $x_1, x_2, \ldots$; in this case the density at x_i is $f(x_i) = P(X = x_i)$ and $P(X \in A) = \sum_{x_i \in A} f(x_i)$ for all events A. X is continuous if its distribution function can be obtained as an integral

$$F(x) = \int_{-\infty}^{x} f(u)\, du$$

of some function $f(x)$, called the density. Of course, in the continuous case, $f(x) = dF(x)/dx$ and $P(X = x) = 0$ for all x. It is common to refer to the distributions of continuous random variables as absolutely continuous distribution functions, however they must be differentiable as well as continuous.

Extension of the concepts of probability distribution and density to two or more random variables is straightforward. The pair of random variables (X, Y) is discrete if at most a denumerable numbers of points $(x_1, y_1), (x_2, y_2), \ldots$ can be obtained.

The joint density at (x, y) is then $f(x, y) = P(X = x, Y = y)$ and

$$P[(X, Y) \in A] = \sum_{(x_i, y_i) \in A} f(x_i, y_i).$$

In particular, the joint distribution function of (X,Y) is

$$F(x, y) = P(X \leq x, Y \leq y) = \sum_{\{i : x_i \leq x, y_i \leq y\}} f(x_i, y_i).$$

The pair (X,Y) is continuous if the joint distribution can be represented as an integral

$$F(x, y) = P(X \leq x, Y \leq y) = \int_{-\infty}^{x} \int_{-\infty}^{y} f(u, v) \, dv \, du$$

of a function f(x,y) called the density.

The **marginal** probability distribution of X is just the probability distribution of X, where it is being emphasized that there are other variables present. Thus, in the discrete case, the marginal density is obtained by summing the joint density over the other variable. In the continuous case, the marginal distribution of X is

$$F_X(x) = P(X \leq x) = F(x, \infty) = \int_{-\infty}^{x} \int_{-\infty}^{\infty} f(u, v) \, dv \, du$$

so that

$$f_X(x) = \int_{-\infty}^{\infty} f(x, v) \, dv$$

is the marginal density of X; the marginal density is obtained by integrating the joint density over the other variable. The conditional density of X, given $Y = y$, is

$$g_{X/y}(x) = f(x, y)/f_Y(y).$$

We write EX for the **expected value** of the random variable X. For a continuous random variable with density f,

$$EX = \int_{-\infty}^{\infty} x f(x) \, dx.$$

The corresponding formula for a discrete random variable with possible values $x_1, x_2, \ldots$ and discrete density f is

$$EX = \sum_{i=1}^{\infty} x_i f(x_i).$$

Conditional expectation of the random variable h(X) is

$$E[h(X)|y] = \int_{-\infty}^{\infty} h(x) g_{X|y}(x) \, dx.$$

It is often convenient to calculate expectation conditionally according to the formula $EY = E\, E(Y|X)$: see, for example, DeGroot (1975, p. 178).

The **variance** of X, written VX, is defined to be $VX = E(X - EX)^2$.

The sets $\{B_1, \ldots B_n\}$ are called a **partition** of S if no pair B_i, B_j have an element in common and $S = B_1 \cup \ldots \cup B_n$.

Theorem 5.2 (Total Probability) Given a partition $\{B_1, \ldots, B_n\}$ of S, then $P(A) = P(A|B_1)P(B_1) + \cdots + P(A|B_n)P(B_n)$.

An elementary result which plays a central role in the foundations of statistics is **Bayes' rule**:

$$P(B|A) = \frac{P(A|B)P(B)}{P(A)}.$$

Given a partition $\{B_1, \ldots, B_n\}$ of S, we may apply the theorem on total probability to the denominator of Bayes' formula to obtain **Bayes' theorem**,

$$P(B_i|A) = \frac{P(A|B_i)P(B_i)}{\sum_{j=1}^{n} P(A|B_j)P(B_j)},$$

for $i = 1, \ldots, n$.

Example 5.1 The problem of false positives.

We provide a simple illustration of the utility of Bayes' theorem. The residents of a community are to be examined for a disease. The examination results are classified as $+$, infection suspected, or as $-$, no indication of infection. But the examination is not infallible. The probability of detecting an infection is only 0.95 and the probability of reporting infection where none exists is 0.01. If 0.2% of the community is diseased, what is the probability of a false positive? We have

$$P \text{ (no infection } |+)$$
$$= \frac{P(+|\text{ no infection}) \cdot P(\text{no infection})}{P(+|\text{no infection})P(\text{no infection}) + P(+|\text{infection})P(\text{infection})}$$
$$= \frac{(0.01)(0.998)}{(0.01)(0.998) + (0.95)(0.002)} \cong 0.84.$$

This probability is undesirably high from a medical point of view, but that is unavoidable since nearly everyone is healthy.

Two random variables X and Y are called **independent** if

$$P[(X \in A) \cap (Y \in B)] = P(X \in A)P(Y \in B)$$

for all sets A and B in the ranges of X and Y, respectively. This is equivalent to requiring that the distributions multiply, $F(x, y) = F_x(x)F_y(y)$, or that the density functions multiply, $f(x, y) = f_X(x)f_Y(y)$. Several random variables $X_1, \ldots, X_n$ are called **independent** if their marginal distributions multiply to yield the joint distributions.

A particular independence model has been central in the evolution of probability theory. This is the **Bernoulli case** of events A_i, $i = 1, 2, \ldots$ which assume equal probability values $P(A_i) = p$ and such that every subset of n of them are independent.

On the basis of the Axioms I–V, V^1 one can prove Borel's famous theorem (see for example, Loéve, 1960, pp. 14–19). Let S_n denote the counting random variable $S_n = I_{A_1} + \cdots + I_{A_n}$ for $n = 1, 2, \ldots$ where I_{A_i} is the indicator function of the set A_i.

Theorem 5.3 [Borel's (1909) strong law of large numbers] In the Bernoulli case,

$$P\left\{\frac{S_n}{n} \to p\right\} = 1.$$

The theorem says that the set of all sample points for which S_n/n eventually comes and remains arbitrarily close to p has probability one; it is commonly interpreted to mean that, whatever probability is, and however independence is interpreted, the ratio S_n/n is likely (in the probability sense considered) to be close to the probability p for large n.

Borel's theorem was the first of many strong laws of large numbers. For a sequence of random variables $X_1, X_2, \ldots$, writing $S_n = X_1 + \cdots + X_n$, we say that the **strong law conclusion** holds for the sequence if $\lim_{n\to\infty}(S_n/n - ES_n/n) = 0$ with probability one. Discussion and proof of the following two strong laws may be found in Gnedenko (1967, §34).

Theorem 5.4 A sequence of independent random variables obeys the strong law if

$$\sum_{n=1}^{\infty} V(X_n/n) < \infty.$$

Theorem 5.5 A sequence of independent and identically distributed random variables, $X_1, X_2, \ldots$, obeys the strong law if and only if their common expectation exists.

Loéve (1960, p. 20) calls a closely related result the central statistical theorem. Let $X_1, X_2, \ldots$ be independent random variables with common distribution $F(x) = P(X_i \leq x)$ and let $F_n(x)$ be the proportion of $X_1, \ldots, X_n$ which do not exceed x; $F_n(x) = S_n(x)/n$ is called the **empirical** distribution function.

Theorem 5.6 (Central statistical theorem)

$$P\left\{\sup_{-\infty < x < \infty} | F_n(x) - F(x)| \to 0\right\} = 1.$$

Borel's result implies $P\{F_n(x) \to F(x)\} = 1$ for each x; the additional content of the above theorem is that the convergence is uniform.

5.2. Interpreting the Axiomatic Theory

The axiomatic approach is to bypass the issue of what probability is and focus instead on how it behaves. In fact, except for the language, the axiomatic theory need not even be about chance or randomness. We emphasize this by providing three examples. First, it is recognized by the use of Venn diagrams as an aid to

TABLE 5.1. Comparison of truth values with probability calculus

$\lvert p \rvert$	$\lvert q \rvert$	$\lvert p \wedge q \rvert$	$\lvert p \vee q \rvert$	$\lvert p \rvert + \lvert q \rvert - \lvert p \wedge q \rvert$	$\lvert p \to q \rvert$	$\lvert p \wedge q \rvert / \lvert p \rvert$
1	1	1	1	1	1	1
1	0	0	1	1	0	0
0	1	0	1	1	1	0/0
0	0	0	0	0	1	0/0

probabilistic intuition that the axiomatic theory applies to areas of subsets of the unit square. Second, as can be seen from Table 5.1, the truth values of the calculus of propositions obey the probability axioms.

Finally, for a probabilistic proof of the important Weirstrass approximation theorem see Iranpour and Chacon (1988, p. 36); the axiomatic theory can be and has been used to prove purely mathematical results.

The axiomatic theory contains an impressive array of results. However, as we have emphasized in Chapter 2, before mathematical results can have practical implications their undefined concepts must be interpreted. What is meant by saying, "the risk of cancer is one in 1,000" or "the probability of rain is 0.3?" Such foundational issues are important and nontrivial. In making the identification of probability with worldly things it is easy to go wrong. For example, in attempting to quantify nuclear safety, the Nuclear Regulatory Commission (1975) lost a lot of credibility by speaking of the probability of a nuclear accident per reactor year. With this formulation, if the probability of an accident per reactor year is one in a thousand then the probability of an accident in two thousand years is two. Expected rate, not probability, per year is the correct tool. Probability is a dimensionless quantity.

Terrence Fine (1973) concludes his thorough mathematical analysis of the foundations of probability on the pessimistic note that none of the various theories are justified. This is the correct conclusion within his adopted framework. His *Theories of Probability* is a needed detailed verification, for the special case of probability theory, of Cramér's reminder that mathematical findings are only conditionally true—i.e., dependent on the truth of their hypotheses—and hence do not have any ultimate validity by themselves. Scientific method is a practice developed to deal with experiments on **nature**. Probability theory is a deductive study of the properties of **models** of such experiments. All of the theorems of probability are results about models of experiments. We can't prove anything about experiments on nature and yet these are the subject matter of scientific method. Something further is needed.

A first step toward interpreting probability is introduction of the undefined concept of a probabilistic experiment. We have seen in Chapter 3 that science is much concerned with approximate permanence in repeated experiments (C,Y).

Chance, is one explanation of this approximate permanence; the rules of probability are thought to apply. It is customary to introduce the concept of a probabilistic experiment or chance situation, by example. Standard examples are flipping a coin and rolling a die, but these are misleading because of their symmetry. Flipping a thumbtack and observing whether it falls point up or point down is a simple intuitive experiment which is not as likely to be misleading. A typical tack will fall

Point up   Point down

FIGURE 5.1. Flipping a thumbtack.

"point up" approximately two thirds of the time (see Fig. 5.1). A similar example more suggestive of the applications is observing whether copies of a valve do or do not shut off the flow of a liquid on demand.

In order to relate Kolmogorov's axioms to probabilistic experiments assume the outcomes of an experiment E partitioned into simple indivisible or elementary events, exactly one of which will occur when E is performed (or observed). Let S be the collection of all elementary events, and α a closed class of subsets of S. Members of the class α are called events and when E is performed the event A ε α is said to occur if a simple event in A occurs; S and its complement Φ are called the certain and impossible events respectively. With this identification of sets with possible outcomes the Axioms and the definition of conditional probability read as before, sets A and B having no element in common being equivalent to events A and B not occurring simultaneously.

A distinction—which must be kept track of—concerning probabilistic experiments is that, like the outcome of Aristotle's sea battle mentioned in Section 1.4, performance changes them from indeterminate to determined. Before a tack is flipped, the point up or down outcome is uncertain, we say it is subject to chance. Flipping resolves this indeterminism, an outcome is realized, and we are no longer in a chance situation. The status of a performed experiment, the outcome of which is not yet known, is usually regarded to be in all respects like an experiment yet to be performed.

Often the outcome of a probabilistic experiment *to be* observed is a random variable Y. Y is not a number; it can only be described in words. We follow the common convention of using capital letters to denote random variables; thus Y is the **potential** outcome of an experiment the response which nature *may* make when the experiment (C,Y) will be performed. After the experiment is performed and the outcome $Y = y$ is observed we have a realization (C,y) of the experiment. The outcome y of a realization—after performance—is then a number.

5.3. Enumerative Probability

The simplest concept of probability—applicable to sample spaces with finitely many elementary events—is that the probability of event A is #A/#S, where #A is the number of elementary events in A and #S is the total number of elementary events. In particular the probability of every simple event is 1/#S; they are equally likely in an enumerative sense. Conversely, any probability with finitely many equally probable elementary events can be calculated according to the rule

$P(A) = \#A/\#S$. This follows since $P(A) = \sum_{s \in A} P(s) = e\#A$ where $e = P(s)$ for $s \in S$ and in particular, $1 = P(s) = e \cdot \#S$. For this reason, enumerative probability is alternatively called equally likely probability.

It is immediate that enumerative probabilities satisfy Kolmogorov's axioms. For example, $\#A \cup B = \#A + \#B$ if A and B are disjoint; dividing by $\#S$, we obtain the addition axiom. The definition of conditional probability is obtained as a theorem—not an axiom or a definition—from the multiplication formula

$$P(A \cap B) = \frac{\#A \cap B}{\#S} = \frac{\#A \cap B}{\#B} \frac{\#B}{\#S} = P(A|B) \cdot P(B).$$

The enumerative theory is the historically first concept of probability (implicitly utilized by Cardano around 1520); it appears in many elementary algebra texts as a motivation for combinatorial analysis, the advanced theory of counting. To calculate the equally likely probability of event C we need only count the number of simple or elementary events in C and the total number of simple events; the probability is then their ratio. A typical example, from the game of poker, is that if all five-card unordered hands are equally likely, then the probability of being dealt four aces is $48/\binom{52}{5}$.

One objection to defining probabilities exclusively in terms of equally likely events, is that many of the most interesting applications cannot be formulated in this way. The equally likely idea of probability may work well for symmetric examples such as poker hands and rolling dice, but in both the valve and the thumbtack examples, it would be exceedingly difficult intuitively to determine mutually exclusive and equally likely simple events.

Even for the simple experiment of flipping coins all intuitions do not agree on what is equally likely. In 1754, the respected mathematician D'Alembert incorrectly calculated the probability of throwing a head in the course of two throws of a coin. He reasoned, that if a head appears on the first throw, the issue is decided, and hence, there are only three cases H, TH, and TT; therefore, the probability is 2/3. More useful answers are 2/4 for exactly one head, and 3/4 for at least one head.

The equally likely concept originates from a time when it was thought necessary to base all mathematics on "self-evident truths"; it is the Euclidean geometry paradigm. But there is nothing self-evident about equally likely for many applied problems, and hence, it would seem that we must look elsewhere for a more flexible concept of probability.

5.4. Frequency Probability

Looking elsewhere, relative frequency presents itself as a promising basis for probability. The **relative frequency** of an attribute A in n observations is the ratio $F_n(A) = S_n(A)/n$ where $S_n(A)$ is the number of observations of A. Note that, since $S_n(S) = n$, relative frequency—like equally likely probability—is the ratio of two quantities obtained by counting; but the countings are done in entirely different and unrelated contexts.

R. vonMises (1957 pp. 28–29) develops a frequency theory along the following lines:

1. It is possible to speak about probabilities only in reference to a properly defined collective.
2. A collective is . . . an unlimited sequence of observations fulfilling the following two conditions: (i) the relative frequencies of particular attributes within the collective tend to fixed limits; (ii) these fixed limits are not affected by any place selection (selection of a subsequence of observations from the collective according to some fixed rule). . .
3. The limiting value of the relative frequency of a given attribute assumed to be independent of any place selection, will be called 'the probability of that attribute within the given collective.'

But the taking of limits here has proved extremely difficult to make precise. The mathematics of limit of a sequence applies to real numbers. Therefore a collective must be a sequence of already realized observations and hence determined. The collective is interesting as a model for what a table of random numbers might mean; for the entry in any given row and column of a specified page of such a table is determined. But we wish to view frequency probability as a property of the indeterminism called chance, and how to do this in terms of a collective seems a little obscure. There are problems with vonMises' frequency theory. In fact, Gnedenko (1967, p. 58) says of vonMises' conditions (i) and (ii) of paragraph 2 above:

The construction of a mathematical theory based on the fulfillment of both these requirements encounters insurmountable difficulties. The fact is that the Principle of Randomness, (ii), is inconsistent with the requirement of the existence of a limit.

A difficulty with defining or interpreting probability as long-term relative frequency is that it is then difficult to see the need or intent of Borel's strong law, our Theorem 5.3. But perhaps we can realize a frequency theory along other lines.

5.5. Propensity Probability—"Tendency To Happen"

An alternative approach to a frequency theory is suggested by some anthropomorphic language of Eisenhart (1963, p. 29); he calls the "limiting mean. . . , the value. . . which each individual measurement. . . is trying to express." Let us recall, from Section 2.4, the character of a formal science. Initially there is a primitive concept which one wishes to investigate. An axiom system is introduced as a model of the concept. Certain key properties of the concept are incorporated as axioms. Initially, the only sense in which the axioms are "true" is that they seem to hold for the concept. The axiom system is studied by purely logical methods to obtain theorems. Conclusions of theorems are only conditional truths; that is, they are true if the axioms are true.

A famous example of a formal science is Maxwell's (1860) dynamical theory of gases. In our Section 3.1 we have quoted from his introduction; he aims to develop—by mathematical argument—an "analogy," the properties of which can be compared with those of gases. It is along these lines that the propensity theory interprets probability. Propensity probability is a formal science to investigate the primitive concept "tendency to happen"; it is a qualified prediction. Certainly we recognize this last interpretation when we refer to a probability of rain as a weather **prediction.** The investigation of "tendencies" is common in physical science. Gravitation, friction, electrical resistance, genetic fitness and heritability—all are "tendencies" when they are applied. Consider the two statements: (i) atoms of radium tend to decay according to a Poisson process (a probabilistic model) and (ii) the tendency for an object to fall in a vacuum is that the gravitational constant is 32 ft/sec^2. We know each of these statements just as surely as the other and for the same reason. Both are approximate models which have evolved in a social process of hypothesis formulation checked by experiment.

In our discussion of science a recurring theme has been the permanence of outcome in repeated experiments. For a long time, only deterministic experiments were considered, where the conditions (causes) determine completely the outcomes (effects). But slowly mankind began to think of a rational interpretation of nature in terms of another kind of permanence which was first observed in games of chance. For some experiments and events the relative frequencies seem to become approximately stable for large n. We have already remarked that a typical thumbtack will fall "point up" approximately two-thirds of the time. By approximately stable we do not only mean for a single sequence of realizations that F_n becomes more or less constant but also that for different sequences the relative frequencies tend to cluster about the same constant.

If it exists, the **propensity probability** P(A,E) of event A for experiment E is the proportion of the time that A tends to happen when E is performed or observed. As Eisenhart suggests, tendency to happen is present in each individual performance but opportunity for expression is very limited.

Probability may be interpreted—

as a measure of an objective propensity—of the strength of the tendency, inherent in the physical situation, to realize the event—to make it happen.

Popper (1983, p. 395)

The manifestation of P(A, E) is, of course, the relative frequency with which A does happen in a series of performances of E. Propensity probability P(A, E) is that which is measured by relative frequency $F_n(A)$. The probability axioms certainly do "seem to hold" for P(A, E) since they hold for $F_n(A)$. Axioms I and II are desirable properties of a chance situation where α is the collection of things which could happen when E is performed and S is the universal outcome. Axiom III is the assumption that a propensity probability exists. Axioms IV, V', and the multiplication rule are all suggested by the corresponding properties for relative frequencies. For example, the identity $F_n(A \cap B) = [S_n(A \cap B)/S_n(A)]F_n(A)$ suggests the multiplication rule for propensity probability since $S_n(A \cap B)/S_n(A)$ is the relative frequency of B in trials where A occurs.

To attach an intuitive meaning to the concept of independence note that if B and C are independent, then

$$
\begin{aligned}
P(B \cap \bar{C}) &= P(B) - P(B \cap C) \\
&= P(B)[1 - p(C)] \\
&= P(B) \cdot P(\bar{C})
\end{aligned}
$$

so that B and $\bar{C}$ are also independent. Now, $P(B|C) = P(B) = P(B|\bar{C})$. The tendency for B to happen is independent (in the grammatical sense) of whether C has or has not occurred; the occurrence of C has nothing to do with the happening of B.

Borel's strong law, our Theorem 5.3, follows from the axioms. Hence, according to the propensity interpretation, the relative frequency of an event in repeated trials approaches its probability with the strongest possible tendency. Propensity probability is derived (not defined) to be the limit of relative frequency in the special Bernoulli case. But propensity probability is only a normal science—a model or analogy—which has no real world validity until it has been checked by experiment.

6

The Fair Betting Utility Interpretation of Probability

6.1. Probability as Personal Price

Consider again the origins of our subject. The formulation of the the de Mere–Pascal–Fermat problem of points was not in terms of equally likely or frequency or personal degree of belief but as a question concerning the value of a player's position in the game. The issue in general is, if one is to receive a prize or payment, but the amount of that payment is uncertain, then how does that uncertainty affect the value of the prize? A modern development is that this question has been answered in terms of a personal price interpretation of probability. This chapter is our version of the details.

Gambling on the outcome of a chance situation E, such as a horse race or a roll of dice, is a major theme in the history of probability and it is in this context that probability as price is most natural. People sometimes object on moral grounds to formulating their affairs in terms of gambling. But life is a gamble though it need not be formulated as a game. Such important commercial activities as insurance and investing are gambling. Many authors, beginning with vonNeumann and Morganstern (1944) and deFinetti (1974, Ch. 3) have viewed probability as price. A survey with appropriate references is Fishburn (1986). Price—which is exchange rate—is an economic concept; hence, it is appropriate to look at their economic ideas from an economic perspective and to make explicit the sometimes unstated economic assumptions. But we strive for an elementary and self-contained treatment which assumes no prior knowledge of economic theory. Varian (1992) is a standard source for economic background and detail.

An essential idea of theoretical economics is that of consumer preference between bundles of k goods. This idea extends to gambling on a chance situation E with possible outcomes $s_1, \ldots, s_k$, by considering x_i, the money available in the case that s_i occurs, to be the ith good. We allow the possibility that x_i might be negative, in which case the consumer assumes a debt in the amount $|x_i|$ if s_i occurs; debt, credit, and accounts with negative balances are facts of life. The gambler may then contemplate his preferences among various random variables X where $X(s_i) = x_i$ for $i = 1, \ldots, k$. The bundle X is the random amount of money available to the consumer when E is performed; X has been called an uncertain prize or a portfolio.

A standard initial economic assumption here is—

A1. The gambler's preferences are weakly ordered by a relation ω which is reflexive and transitive. Indifference ι and preference ρ are then defined in terms of ω: $X \iota Y$ if and only if $X\omega Y \wedge Y\omega X$ and $X\rho Y$ if and only if $X\omega Y \wedge \sim Y\omega X$. The relation ι is an equivalence and ρ is asymmetric and transitive. A further entirely reasonable assumption is that more money is preferred to less.

A2. If $X \geq Y$ but $X \neq Y$, then $X\rho Y$. A constant portfolio c is worth c regardless of how E turns out; it is sure money. Hence, if $X\iota c$, then c is the amount of sure money which an individual prefers equally to the uncertain prize X. Note that, by A2, for constant portfolios c and d, $c\omega d$ is equivalent to $c \geq d$.

Definition If $X \iota c$ then c is a personal economic **value** of X and if $X \iota v(X)$ for all X, then v is a **value function**.

Definition A real-valued function $u(X)$ is a **utility** representation of ω if $X\omega Y$ when and only when $u(X) \geq u(Y)$.

Clearly, if a utility exists then all portfolios are ω comparable since their utilities can be ordered.

Theorem 6.1 If A1, A2 and v is a value function, then (i) $v(X)$ is a utility representation of ω and (ii) $v(X)$ is the unique personal value of X.

Proof
To establish (i), first assume $v(X) \geq v(Y)$. Then $X\iota v(X)\omega v(Y)\iota Y$ and $X\omega Y$. Conversely, if $X\omega Y$ then $v(X)\iota X\omega Y\iota v(Y)$, and $v(X)\omega v(Y)$. Therefore v is a utility. To prove (ii), suppose $X\iota v_1$ and $X\iota v_2$. From A_1 we have $v_2\omega X\omega v_1 \wedge v_1\omega X\omega v_2$ and from A_2, $v_2 \geq v_1 \wedge v_1 \geq v_2$ or $v_1 = v_2$.

A basic mechanism for gambling is that two individuals, wishing to bet on A and $\bar{A}$ (the complement of A), respectively, place agreed amounts c and d in a pot and receive in return the common amount $c + d$ of A and $\bar{A}$ tickets. Here c and d, called the stakes, are non-negative and $c + d$ is positive. The significance of an amount q of A tickets is that a pot of amount q has been set aside until E is performed; it entitles the bearer to receive the entire pot or nothing according as A does or does not occur. We may view the gambling mechanism as a transaction in which two gamblers have purchased amount $c + d$ of A and $\bar{A}$ tickets at costs of c and d or unit prices of $c/(c + d)$ and $d/(c + d)$.

Bets are offered in terms of odds. If a gambler offers c to d odds on A, he means that he will purchase any quantity of A tickets at unit price $c/(c + d)$. Offering c to d odds against A expresses willingness to take the "other side" of the above bet. That is purchase any quantity of $\bar{A}$ tickets at unit price $d/(c + d)$.[1] A gambler considers a bet, at given odds, to be **fair** if he will take either side.

---

[1] In practice, the size of bet which a gambler will accept is limited by available funds and the possibility of ruin. We return to this point later in this chapter and in the next.

The characteristic assumption concerning fair bets is that they exist. Preference between bets translates into preference between portfolios. If a gambler bets at odds of c to d on A, then his initial endowment X becomes $X + q(I_A - p)$, where q is the size of the pot and $p = c/(c + d)$. Likewise betting q, at odds of c to d against A, changes initial endowment X to $X + q(I_{\bar{A}} - \bar{p})$ where $\bar{p} = 1 - p$. Therefore, if he will offer odds of c to d both on and against A, then $X + q(I_A - p) \; \omega X$ and $X - q(I_A - p) = X + q(I_{\bar{A}} - \bar{p})\omega X$ for all X and q. In the second equation, substituting $X + q(I_A - p)$ for X, we get $X \omega X + q(I_A - p)$. The fair bet assumption is equivalent to—

A3. For each A there is a **fair price** prA, $0 \le prA \le 1$, such that

$$X \iota X + q(I_A - prA) \text{ for all X and q.} \qquad (6.1)$$

A set function satisfying Kolmogorov's axioms (Section 5.1) is called a prob ability.

Theorem 6.2 (The fundamental theorem of personal price probability) A system of prices is fair if and only if it is a probability and economic value is expected value for that probability.

Theorem 6.2 provides special insight which is missing in other theories of prob-ability. A modern tendency, concerning the expected value of a random variable, is to emphasize "expected" and deemphasize "value." Indeed, expected value is sometimes shortened to expectation. From the present point of view this is regret-table. We have all heard jokes about "expecting" a man to have fewer than two arms and a family to have a nonintegral number of children. Actually, one "expects" that, when a die is rolled, 3.5 dots will never show on the up face. The above theorem provides motivation—which is completely bypassed in the axiomatic theory—for the concept of expected value; it is not so much "expected" as it is "value." One interpretation of probability is risk neutral personal unit price and expected value is the corresponding economic value.

Theorem 6.2 is a substantial simplification since it allows us to compare portfo lios and determine fair price probability by comparing expectations, which are real numbers. For example, consider the following.

Corollary 6.1 Given two bets (I) risking c dollars if $\bar{A}$ occurs to win d if A occurs and (II) risking d if A occurs to win c if $\bar{A}$ occurs then (I) is weakly preferred to (II) if $pr A \ge c/(c + d)$. The two bets will be preferred equally, or "fair" if and only if prA $= c/(c + d)$. Proof of the corollary consists of observing that $E(X + dI_A - cI_{\bar{A}}) \ge E(X + cI_{\bar{A}} - dI_A)$ where X is prebet endowment, reduces to prA $\ge c/(c + d)$.

We may call A3 the assumption of risk neutrality since, from Corollary 6.1, it expresses the idea that betting behavior is independent of prebet endowment and bet size.

From Corollary 6.1, the following are equivalent: (i) c to d fair odds on event A, (ii) $\tau \, pr A = c/(c + d)$, and (iii) Pr A/$pr \bar{A} = c/d$. Hence, we may define odds A = pr A/$pr \bar{A}$; then pr A = odds A/(1 + odds A).

As an example, let us solve the problem of deMere with which we began Chapter 5. A game between two players is won by scoring three points. How should the pot be divided if play is discontinued prematurely when the score is two points to one? It was Pascal who first gave the correct answer, that if their chances of winning a single point are equal (and assuming independence), the pot should be divided in the ratio one to three. If play were continued, the possible final scores would be s_1 = three to one, s_2 = three to two, and s_3 = two to three with fair betting probabilities of $1/2$, $1/4$, and $1/4$. Writing $A = \{s_1, s_2\}$ then the first player can be considered to be holding a portfolio q I_A, where q is the size of the pot. The fair value of such a ticket is $EqI_A = 3 q/4$. Similarly, the fair value of the second player's position in the game is $EqI_{\bar{A}} = q/4$. This is Pascal's answer.

Proof of Theorem 6.2
First assume $X \iota EX = \sum x_i P\{s_i\}$ where PA is some probability. From Theorem 6.1, EX is a utility representation of ω so that $X \iota Y$ if and only if EX = EY. Therefore, since $EX = E[X + q(I_A - PA)]$, prA = PA is a solution of (6.1) for every X, q and A.

To prove the converse we first show that $v(X) = \sum_{i=1}^{k} x_i p_i$ is a value function where $x_i = X(s_i)$ and $p_i = pr(s_i)$, for $i = 1, \ldots, k$. Applying (6.1) repeatedly, we obtain $X \iota X - x_1(I_1 - p_1) - x_2(I_2 - p_2) - \cdots - x_k(I_k - p_k) = \sum_{1}^{k} x_i p_i$ where $I_i(s)$ is 1 if $s = s_i$ and is 0 otherwise. Hence, from Theorem 6.1, v(X) is a utility and the unique value.

Expression (6.1) implies $v(X) = v[X + q(I_i - p_i)] = v(X) + q(p_i - p_i \sum_{j=1}^{k} p_j)$, so that either $p_i = o$ for all i or $\sum_{j=1}^{k} p_j = 1$. The first degenerate solution denies A2 so that we are left with $\sum_{j=1}^{k} p_j = 1$. From $V(X) = v[X + q(I_A - prA)]$, we obtain $prA = \sum_A p_i$.

It is now straightforward that prA is a probability. Kolmogorov's Axioms I and II of Section 5.1 hold with $S = \{s_1, \ldots s_k\}$ and α the class of all subsets of S. Axiom III is part of A3. The probability addition property for disjoint sets A and B follows from $pr(A \cup B) = \sum_{A \cup B} p_i = \sum_A p_i + \sum_B p_i = prA + prB$. And axiom 4 is a consequence of $pr S = \sum_{i=i}^{k} p_i = 1$. Finally, $v(X) = \sum_{i=1}^{k} x_i p_i = EX$. The expected value with respect to the probability prA.

An alternative way of looking at the risk neutrality assumption is in terms of additive value. A value function is **additive** if the value of two portfolios is the sum of their values. A consequence of Theorem 6.2 is that—in the presence of A1 and A2, A3 necessarily implies the existence of an additive value function. Additivity of value is, for practical purposes, also a sufficient condition that A3 should hold. Suppose A1, A2 and that v(X) is additive. Except for pathologies, Aczél (1966), v(X) is of the form $v(X) = C \cdot X = \sum_{i=1}^{k} c_i x_i$, where $c_i = v(I_{\{s_i\}})$ From Theorem 6.1, $C \cdot X$ is a utility so we may solve (6.1) for prA by solving $C \cdot X = C[X + q(I_A - prA)]$ to obtain $prA = \sum_{s_i \in A} c_i / \sum_{i=1}^{k} c_i$. From A2, $I_{\{s_i\}} \rho 0$ and hence $c_i > v(0) = 0$. Therefore, A3 holds.

6.2. Probability as Personal Degree of Belief

This section verifies an issue usually taken for granted: that fair betting probability measures degree of belief. Theorem 6.2 tells us only that probability may be interpreted as risk neutral personal price.

The fair bet theory models much gambling practice and yet—as deFinetti was aware—the assumption A3 of risk neutrality is at odds with economic principles; risk aversion is considered to be prudent business practice. Fair betting does not assign a **realistic** utility interpretation to probability. Initially this looks like a defect in the theory. But we aren't really trying to provide a realistic model of economic decision-making; that is the purpose of Nau and McCardle (1991). Instead, the main use of the fair betting argument has been to justify a degree of belief interpretation of probability.

Realistic betting behavior will reflect both attitude toward risk and attitude toward uncertainly. But to conceptually get at an individual's degree of belief, which is attitude toward uncertainty independent of attitude toward risk, we need to consider personal preferences and prices determined *as if* the person were risk-neutral. In order to quantify personal degree of belief, it is proper counterfactually to consider risk-neutral preferences. We need an additional assumption.

A4. Degree of belief about how an experiment will turn out is to be quantified by risk neutral gambling preference.

The unrealistic nature of probability interpreted as price explains why it has proved difficult to measure. Kadane and Winkler (1988) and Schervish et al. (1990) discuss the difficulties inherent in measurement or elicitation. Kadane and Winkler calculate that for three common methods, elicited probability will not equal true personal probability if the utility function is nonlinear. The paper of Schervish et al. is also concerned with difficulties of separating personal probability from utility on the basis of observed preference. Concerning these elicitation difficulties, Nau and McCardle (1991) take the view that the separation of belief from risk preference "is inessential to the characterization of economic rationality in terms of observable behavior." But such a separation *would* be essential to the interpretation of probability as degree of belief.

The personal degree of belief theory is the logic of a rational economic person; discounting attitude toward risk, when betting on the outcome of an experiment he will put his money where his belief is. Note that the concern is an experiment which has an outcome. We must be dealing with an experiment in the sense of Section 5.2, that is, a recipe for performance of the form (C,Y). For if we ask, "Betting on what?" or "Degree of belief about what?" then the answer to either question is, "How a chance experiment will turn out." Moreover, the experiment must eventuate. A person would be foolish to wager funds on the truth of a proposition when there is no possibility of payoff. It isn't rational to put money in a bank which is in default.

But people sometimes have beliefs, which they think are rational, when there is no chance experiment whose outcome we can bet on. An important example concerns probability statements about the truth of scientific theories. A current proposal is to base scientific method on coherence; see Berger and Berry (1988) and Howson and Urbach (1989). But where is the chance experiment which determines the gambling scientist's payoff? At what stage and in what sense does the experiment eventuate? Degree of belief is sometimes difficult to formulate in terms of a gambling situation. Our discussion of probabilities may not extend to such cases. An essential part of the fair betting probability argument, and its extension to degree of belief, is that we are concerned with a chance experiment which does eventuate.

6.3. Conditional Subjective Probability

Of course no interpretation of probability will be complete without some discussion of conditioning. We use Corollary 6.1 to provide a proof, in informal outline, that fair prices satisfy the multiplication rule of probability. The price interpretation is that prA is a gambler's (prior) unit price for an A ticket on an experiment before learning anything of its outcome. Conditional probability pr(B|A) is his (posterior) price per unit for a B ticket on the experiment upon learning only that A happens. Suppose prA = 0.6 and pr(B|A) = 0.5. The gambler is prepared to buy (on account) (i) a 50-cent A ticket for 30 cents and (ii) when and if he learns only that A happens a one dollar B ticket for 50 cents. Making these purchases the histories of his accounts would appear as in Table 6.1.

From the last column we see that he would have in effect purchased a $1 A ∩ B ticket for 30 cents; so he must be prepared to make this purchase. Considering the "other sides" of these bets he is willing to commit to the purchases of Table 6.2.

TABLE 6.1. Account history scenarios

	Prior	Posterior		Maturity		
Outcome	A	A	B|A	A	B|A	Total
A ∩ B	−30	20	−50	20	50	70
A ∩ B̄	−30	20	−50	20	−50	−30
Ā ∩ B	−30	−30		−30		−30
Ā ∩ B̄	−30	−30		−30		−30

TABLE 6.2. The "other side" of Table 6.1

	Prior	Posterior		Maturity		
Outcome	Ā	Ā	B̄|A	Ā	B̄|A	Total
A ∩ B	−20	−20	−50	−20	−50	−70
A ∩ B̄	−20	−20	−50	−20	50	30
Ā ∩ B	−20	30		30		30
Ā ∩ B̄	−20	30		30		30

So he is, in effect, willing to purchase a $1 $A \bar{\cap} B$ ticket for 70 cents. This is the "other side" of the above 30-cent bet on $A \cap B$. Therefore his $pr(A \cap B)$ exists and equals 0.3, which is (0.5)(0.6) dollars. In general $pr(A \cap B) = pr(B|A) \, pr(A)$.

This rounds out the argument that Kolmogorov's axiomatic theory of probability may be interpreted as individual risk-neutral price.

7

Attitudes Toward Chance

7.1. Indeterminism and Chance

The idea of a probabilistic experiment or chance situation was introduced by example in Section 5.2. There are several attitudes concerning the indeterminism which we call chance: (i) chance is the opposite of determinism; (ii) the situation is in principle completely determinable but the initial conditions have been incompletely specified; (iii) in some respects the world is fundamentally random; and (iv) the indeterminism is located in the mind(s) of the observer(s) rather than in the exterior world. We shall immediately discuss (i) and (ii). The third attitude may be correct, but it begs for an explanation which we cannot provide. Discussion of attitude (iv) was begun in Chapter 6 and will continue.

The first attitude is rarely stated but frequently followed by statisticians of all persuasions. But chance is not the only alternative to determinism. The outcome of an experiment need not be subject to the rules of probability just because it isn't uniquely determined by the initial conditions; there is an additional requirement of statistical control. Eisenhart (1963, Section 3.1) traces this observation to many famous scientists: Galileo, Millikan, Shewhart, and Student, But perhaps Deming (1986, p. 312) is most explicit:

A first lesson in application of statistical theory. Courses in statistics often commence with study of distributions and comparison of distributions. Students are not warned in classes nor in the books that (statistical methods) serve no useful purpose . . . unless the data were produced in a state of statistical control. The first step in the examination of data is accordingly to question the state of statistical control that produced the data.

And earlier, Deming (1950, pp. 502–3) states—

In applying statistical theory, the main consideration is not what the shape of the universe is but whether there is any universe at all. No universe can be assumed, nor . . . statistical theory . . . applied unless the observations show statistical control.

. . . Very often the experimenter, instead of rushing in to apply [statistical methods] should be more concerned about attaining statistical control and asking himself whether any predictions at all (the only purpose of his experiment), by statistical theory or otherwise, can be made.

By simple statistical control, these writers mean that we are sampling from a population or universe, that successive performances of the experiment are independent and subject to the same law of probability. Indeed, most of the sampling theory of statistics is about independent and identically distributed (i.i.d.) random variables. Of course, statistical inference is possible for other probability structures, but the point here is that experimental results may be haphazard rather than random and initial adjustment of the experimental procedure and testing of the results will be necessary before the rules of probability can be applied with any confidence. Several illustrations of how experimental results may deviate from intuitive distributions will be presented in Examples 7.2–4.

Attitude (ii) is the classical physical concept of determinism and chance, in which prediction plays a central role. The classical concept of determinism visualizes a sequence of sources of error which, if understood and controlled or accounted for, would improve predictive accuracy until, in the limit, experimental outcome would be predicted exactly. The explanation of chance is then, that the world is deterministic but chance results from failure to account for all initial conditions with sufficient precision.

A first example is that conceivably we could learn to control the flip of a coin, obtaining heads or tails at will; chance variation is the consequence of not having built a machine which controls the initial conditions sufficiently accurately.

For a more extensive illustration, reconsider Example 3.1, the ballistics example of shooting at a target. The author is familiar with some of the data from this much-studied experiment. The predictions of the parabolic theory will be in error; successive firing will result in an ellipsoidal pattern of impacts more or less near the target. A first reason is that the parabolic trajectory ignores air resistance. Adjusting for air resistance improves prediction and yields a new ellipsoidal pattern of impacts nearer to the target. Further sources of error are (i) humidity (ii) curvature of the trajectory due to spin of the projectile (iii) wind, both average velocity and gusting, (iv) amount of projectile propellant (v) earth curvature, and (vi) the coriolis effect of the earth spinning under a projectile in flight. According to the classical explanation, chance is due to not having identified all of the relevant factors or learned to control them with sufficient precision.

When we follow the recipe (C,Y), there will be additional conditions, U, not monitored or controlled and perhaps unknown which may also affect the outcome; the actual operating mechanism, will be (C,U,Y). The classical explanation is that the chance behavior of Y is due to variation in U. Repetition of (C,Y) is in fact a sequence (C,U_1,Y), $(C,U_2,Y) \ldots$, and (C,Y) is the class of all repetitions which might be performed using the given recipe. Note that this is really an explanation of indeterminism not chance—the probability axioms have not been justified.

7.2. The Location of Probability—Mind or Matter?

The most important issue, for the meaning and use of a probability, is whether it is subjective or objective. Bruno deFinetti (1974, p. x) settles the issue neatly by stating that all probabilities are subjective. He has a point.

First, we dispose of a linguistic problem. One meaning of the word *subjective* is biased or prejudiced. That is not the relevant meaning here. Here, objective signifies "of, or having to do with an object" as distinguished from something existing only in the mind of the subject or person thinking.

Now let us proceed to deFinetti's comment; it is also true of other basic scientific quantities such as length, time, coefficient of friction, etc. His point is more general, and is often credited to Immanuel Kant. At the epistemological level it has to be admitted that knowledge is influenced by perceiving and structured by mind. It is an important observation and seems absolutely true. All mathematical models and therefore all probabilities and all statistical analyses are to some extent constructs of the mind. This state of affairs has caused some, for example, Berger and Berry (1988), to become skeptical of objectivity and to abandon the search for truth about the world in favor of understanding the state of ones own mind.

But, if probabilities are just one man's opinion (personal) then why should science as a whole pay any particular attention to them? Scientists are used to thinking that they function at a physical or ontological level, that they are studying and discovering facts about nature which are independent of any observer. Nature is what is left after the observer dies. Ultimately nature is unknowable but aspects of nature such as length, time and probability exist and can be understood in the sense of trustworthy belief through a process of logic checked by experiment. This way of thinking has been enormously successful and it is hard to argue with success.

All probabilities are conjectural since they are based on models and assumptions; they are only as good as those assumptions. But all probabilities are not subjective to the same extent. The key here is suggested by considering the two statements: "This is a fair coin" and "I think this is a fair coin." The first would be checked by flipping the coin, the second by offering me various bets to determine the odds which I would accept.

The test of objectivity is the manner of checking. If the correctness of a probability has been evaluated by performing an experiment on nature then presumably that probability has to do with the object of that experiment; it may be conjectural or even wrong but it is about nature. On the other hand if a probability cannot be checked or is to be checked by introspection or observing the behavior of the person thinking then it is subjective. An important possibility is that probabilities which are initially only subjective may make the transition to objectivity through a process of experimental verification.

To what extent are probabilities properties of nature and may we think of them as physical or objective? Where is the location of probability? These are important questions. The answers will depend on the concept of probability.

7.3. Equally Likely

The equally likely concept is a good place to start. A recurring justification of the assignment of equal probabilities has been the principle of **insufficient reason**—that if there are no grounds for assigning unequal probabilities then equal probabilities

TABLE 7.1. Winnings from various bets

Event\bet	I	II	III	I + II + III
1	b	−a	−a	b − 2a
2	−a	b	−a	b − 2a
3	−a	−a	B	b − 2a

should be assigned. Equally likely probabilities arrived at through the principle of insufficient reason will be called **Laplacian**, for Laplace made much use of the principle. Apparent symmetry or invariance, as with dice, is the usual reason for assigning Laplacian probabilities.

Let E be the statement that there are grounds for assigning equal probabilities and U that there are grounds for assigning unequal probabilities. The principle of insufficient reason is to act like not U implies E. This is the fallacy of appeal to ignorance; it is logically incorrect if probability is located beyond mind, for we may have no information at all and hence no grounds for any assumption about nature. Then, in particular, not E and not U will be true so that if not U implies E then E and not E are true, a contradiction.

But while lack of knowledge is grounds for nothing in nature it does seem to be grounds for knowing ones own mind. Thus Laplacian probability should be viewed as subjective, indicating a state of individual complete lack of knowledge. This conclusion is supported by the following argument which shows that gambling probabilities corresponding to no information are Laplacian.

Consider three elementary events none more reasonably expected to occur than another.

The bets I, II, and III of Table 7.1 are bets on events 1, 2, and 3 respectively, at odds of a to b. Because of the invariance of the situation under permutations of the elementary events, if I is fair, then II and III will similarly be fair. Then I + II + III is a fair bet with constant winnings b − 2a. Therefore, if I is a fair bet, then b − 2a = 0 and the gambling probability of event I is $a/(a + b) = 1/3$. Each of the three events will be assigned this same probability.

Let us apply the acid test. How will a Laplacian probability be checked? By asking questions or testing the behavior of the Laplacian: What might make the probabilities unequal? You have no hint of which event will occur? None of the events can more reasonably be expected to occur? Laplacian probabilities are subjective.

Example 7.1 The prisoner's dilemma of three prisoners, call them A, B, and C, two have been chosen to be executed. Prisoner A asks their jailor to name one prisoner other than himself who will be executed. The jailor responds that B will die. Now A knows that either he or C will live; hence he reasons that his probability of living has been increased from 1/3 to 1/2 just by getting the jailor to answer his question. Can this be?

Since the jailor responds after the prisoner to live is chosen, nothing the jailor says can alter the tendency that A **will** live. So in this sense, A's reasoning must be in error. Here, as deFinetti says, probabilities do not attach to objects. The phrasing

TABLE 7.2. Joint probabilities

Jailor's response	Prisoner to live		
	A	B	C
B	p/3	0	1/3
C	q/3	1/3	0
	1/3	1/3	1/3

of the problem suggests that the probabilities 1/3 and 1/2 are Laplacian and hence subjective. There is no experiment in nature which will check their correctness.

One may question the principle of insufficient reason but, applied here, it does yield the equal probabilities of 1/3. However, in calculating the probability that A lives given the jailor's response, we will see that there are "grounds for assigning unequal probabilities" to the two remaining possibilities.

We analyze only the situation that the jailor does not lie. He has two possible responses and there are three a priori choices of which prisoner will live, a total of six possible outcomes to the joint experiment. Note that the conditions for the existence of fair bet probability—with which we concluded Section 6.2—are met; here we are concerned with a conceptual chance experiment which does eventuate. Joint probabilities appear in Table 7.2.

We may calculate from first principles that $q = 1 - p$ and p is the conditional probability of response B given that A lives. The probability that A lives given response B is

$$p/(1 + p) \tag{7.1}$$

Whether A's reasoning is correct depends on how the jailor's response is interpreted. If we suppose that the jailor adopts the equally likely strategy, $p = 1/2$, then P(A lives |response B) = 1/3 while P(C lives | response B) = 2/3 and A's Laplacian reasoning is incorrect since the two events are not equally likely. But if $p = 1$ then the required probability *is* 1/2. In any event the expression (7.1) has absolutely nothing to do with whether A *will* live; it is a purely subjective probability.

The solution is sensitive to particular language; suppose that A had asked, "Is B to die?" and the jailor had said "yes." Now, with equal a priori probabilities, the probability that A lives given the jailor's response is P(A lives)/P(B dies) = (1/3)/(2/3) = 1/2, which suggests that A's Laplacian computation may be correct. That however, is the solution to a different problem where we are given that B is to die. For the problem as originally stated we are given that the jailor says truthfully that B is to die. The latter event implies the former but not vice versa.

If, after hearing the jailor's response, we allow A to trade places with the remaining prisoner, then we have a problem which has received much public attention, Morgan et al. (1991). From Table 7.3, the unconditional probability that A lives by switching is 2/3, while the probability that he lives by not switching is p/3 + q/3 = 1/3. It appears that A can boost his probability of living from 1/3 to 2/3 by always switching regardless of the jailers response. This is correct but it must be

TABLE 7.3. The prisoner's dilemma

Jailor's response	Prisoner to live a priori	Probability	Outcome Switches	Outcome Doesn't switch
B	A	p/3	d*	l
B	B	0	d	d
B	C	1/3	l	d
C	A	q/3	d	l
C	B	1/3	l	d
C	C	0	d	d

*l = A lives, d = A dies.

remembered that these are subjective probabilities; A can boost what he thinks his probability of living is from 1/3 to 2/3 by always switching.

Conditionally, P(A lives by switching | R = B) = $\frac{1/3}{1/3+p/3}$ = $\frac{1}{1+p}$ and P(A lives by not switching | response B) = $\frac{p/3}{p/3+1/3}$ = $\frac{p}{1+p}$ ≤ $\frac{1}{1+p}$ with equality only when p = 1. A still thinks he should switch but his probability of living is not always 2/3, it depends on p.

The above conditional and unconditional solutions are consistent with one another since P(A lives by switching) = P(A lives by switching | response B) P(response B) + P(A lives by switching |response C) P(response C) = $\frac{1}{1+p}\frac{1+p}{3}$ + $\frac{1}{1+q}\frac{1+q}{3}$ = 2/3.

The primary interest of this problem is that the solution and its meaning are strongly dependent on additional assumptions and interpretation. It is particularly difficult to remember that A does not change his objective chances of living by switching, only what he thinks his chances are.

An experiment to which the probability (7.1) corresponds, is one in which the execution is "called off" if the jailor does not respond B. Consider the "trial" suggested by Table 7.2. Repeat this trial independently, calling off execution till response B is obtained; then note whether A lives. The probability that A lives in the first trial for which response B is obtained is then

$$\sum_{i=1}^{\infty} \left(\frac{2-p}{3}\right)^{i-1} p/3 = \frac{p}{1+p} = P(A \text{ lives}|\text{response B}).$$

The last computation is interesting since conditional probabilities may be similarly developed in general, in terms of independent trials rather than the usual reverse development. Suppose that A and B are events of experiments E. Consider the conditional experiment in which E is performed independently until B occurs. The probability that A occurs in the first trial in which B occurs is

$$\sum_{i=1}^{\infty} [1 - pr(B)]^{i-1} pr(A \cap B) = pr(A \cap B)/pr(B) = pr(A|B).$$

This concludes our discussion of the prisoner's dilemma.

Not all equally likely probabilities will be Laplacian. We may conjecture that a situation E has the property that when it occurs the elementary outcomes tend to happen equally often, that is, a uniform propensity probability exists. If the conjecture survives experimental test, as for some genetic situations and with Bose-Einstein physics, then it becomes science. Or we may **arrange** it so that probabilities are equal, as when a table of random numbers or computer program previously tested for patterns is used to choose a sample from a finite population. Such probabilities are physical and objective, they are conjectural properties of nature which can be checked by experiment.

But, remember that such physical equally likely probabilities are **only** conjectural. Symmetries are sometimes only apparent and it is even difficult to **build** a mechanism which will generate equal probabilities. While a die may appear symmetric and consequently we may assign the faces equal probabilities, there is no guarantee that approximately equal frequencies **will** be obtained when the die is rolled. This seemingly obvious point can sometimes be a source of confusion.

Example 7.2 That the sexes are equally likely at birth is a good first approximation, but when the matter is examined more carefully it is clear that a male birth is slightly more likely. The question has an ancient history, some of which may be found in Todhunter (1949) indexed under the heading "Births of boys and girls."

Example 7.3 Public confidence in the fairness of the United States military draft was undermined by apparent non-randomness of the 1970 draft lottery. The 366! Possible orders of birthdays appear not to have been equally likely. Fienberg (1971, p. 255) summarizes: "Randomization is not easily achieved by the mixing of capsules in a bowl."

Example 7.4 Deming (1986, pp. 351–2) discusses a sampling experiment in which a "lot" of 50 beads is drawn mechanically by dipping a paddle with 50 depressions in it into a box containing 3,000 white and 750 red beads.

Cumulated average. Question: As 20 per cent of the beads in the box are red, what do you think would be the cumulated average, the statistical limit, as we continue to produce lots by the same process over many days?

The answer that comes forth spontaneously from the audience is that it must be 10 because 10 is 20 percent of 50, the size of a lot. Wrong.

We have no basis for such a statement. As a matter of fact, the cumulated average for paddle No. 2 over many experiments in the past has settled down to 9.4 red beads per lot of 50. Paddle No. 1, used for 30 years, shows an average of 11.3.

The paddle is an important piece of information about the process. Would the reader have thought so prior to these figures?

7.4. Frequency and the Law of Large Numbers

The author has observed what appear to be instances of stable frequencies. For example, if an ordinary thumbtack is tossed against a vertical backboard and allowed to come to rest on a hard horizontal surface, then the frequencies of "point up" and "point down" seem to display approximate stability. It seems that long-run stability of relative frequency should somehow provide an objective interpretation of probability but here too there are problems. The difficulties with vonMises' direct taking of limits in a collective have already been mentioned in Section 5.4. We discuss three further approaches: (i) proof from Borel's theorem, (ii) empirical justification, and (iii) propensity probability.

Borel's theorem is a consequence of the axioms of probability and, hence, for any interpretation, the relative frequency of Bernoulli events will almost surely approach their common probability. But Borel's theorem does not prove that any result *will* hold for some series of real world experiments. Like all mathematics, Borel's theorem is not about the real world, it is about what has been postulated by thought. Probability is at most a mathematical model of the world and the hypotheses of Borel's theorem are difficult to justify and the conclusion is, from a practical perspective, imprecise. The hypotheses of independence and countable additivity are particularly difficult to justify. Independence will be suspect since ultimately everything is related to everything else. Justification of independence then must take the form that relatedness makes little difference for the purpose at hand.

Credibility of countable additivity—V' of Chapter 5—is enhanced by its equivalence to the continuity property, Theorem 5.1. The utility of infinite models in mathematics is that they serve as approximations for large finite situations. In particular, continuous probabilities will be for the purpose of approximating the discrete probabilities of finite sample spaces with many points. From this purpose and from Theorem 5.1 we see the motivation for requiring countable additivity (V'). But this motivation is mostly wishful thinking about mathematical neatness.

The conclusion of Borel's strong law, that almost surely relative frequency of an event eventually comes and remains arbitrarily close to the probability of the event, is imprecise in two respects. First, "eventually" has no practical meaning: how large must "n" be before F_n will be a specific distance from p? Second, what is the practical meaning of almost surely?

Borel's strong law is a purely mathematical result which must be interpreted and supported in some way before it can be of any practical consequence. We do learn about nature by augmenting and checking mathematical theory with experimentation. Cramér (1946, p. 144) writes

We may regard it as an established empirical fact that the "long-run stability" of frequency ratios is a general characteristic of random experiments, performed under uniform conditions.

But we have seen in Sections 7.1 and 7.3 that Deming and others caution against such an attitude. How shall we recognize "random experiments, performed under uniform conditions?"

How might we check the long run stability of relative frequency? If we are to compare mathematical theory with experiment then only finite sequences can be observed. But for the Bernoulli case, the event that frequency approaches probability is stochastically independent of any sequence of finite length. For, suppose A to be any event involving only the first n trails and B is the event that frequency approaches probability. From Borel's strong law, Theorem 5.3, we have $P(B) = 1$ and $0 \leq P(A \cap \bar{B}) \leq P(\bar{B}) = 0$. Therefore $P(A \cap B) = P(A) - P(A \cap \bar{B}) = P(A) \cdot 1 = P(A) \cdot P(B)$. Any event A which is observable, is independent of—has nothing to do with—Borel's event B. Long-run stability of relative frequency cannot be checked experimentally. There are neither theoretical nor empirical guarantees that, a priori, one can recognize experiments performed under uniform conditions and that under these circumstances one *will* obtain stable frequencies.

Turning to propensity probability, we find a frequency theory which is consistent with the evolutionary view of science developed in Chapter 3. Propensity probability or "tendency to happen" is that property of a chance situation (or recipe) which is estimated by relative frequency. That such a property exists is conjectural; but its existence in the sense of trustworthy belief can be tested. Not, to be sure by checking the existence of a limit but by comparing predictions which *can* be checked (always subject to the possibility of error) with finite experimental results. Much of the sampling theory of statistics is about how to do this. Stable relative frequency is a theoretical consequence of conjectural axioms, via Borel's theorem; it does not serve as a suitable starting point and it can't be checked directly.

The conclusions of the theorems of a formal science such as propensity probability are only conditional truths; they are true if the axioms are true. The real world validity of the whole must be judged by comparing theoretical prediction with actual experiment. For a few, all too few, applications this has been done and documented so that the formal science of propensity probability *for those applications* has been advanced to the status of trustworthy conclusion, which is science. However, for other chance experiments, where belief has not been tested, then propensity probability is just formal science; the probability model "though consistent with itself" may simply not be an appropriate "analogy."

7.5. The Single Instance

Frequency probabilities, particularly of the vonMises type, have been criticized on the grounds that they do not say very much about a single observation; since probability is a property of a collective, we cannot speak of the probability of "heads" on a single flip of a coin. Equally likely and fair bet probabilities on the other hand, make no reference to embedding a particular flip in a class and hence are applicable to the single instance.

Now turning to propensity probability, since it is a property of experimental situation E, that is the class of all trials performable using that recipe, trials of the class play a symmetric role in the definition. Therefore it is merely a convention whether probability is a property of the class or a common property of each trial in the class. Since flipping a coin is a repeatable situation, we can speak of the probability of "heads" on a single flip even though it would take many flips to accurately estimate that probability. Frequency does not *directly* check probability in the single instance but indirect verification is the rule for scientific theories.

For example, the definition and measurement of the coefficient of friction of a mechanical system depends on much theory and many experiments previously performed on other mechanical systems.

We think that frequency probabilities, particularly those of the propensity type, *are* applicable to the single instance but that the nature of "the single instance" needs careful discussion. Propensity probability is a property of an experimental recipe but it "rubs off" on the individual performances of that recipe. In speaking of the propensity probability of "heads" the reference is not to the "coin" but to the "flip of the coin." Here much is usually understood about the manner of "flipping the coin;" if the coin is bent or the flipping is done by machine then the tendency for "heads" to occur may be changed.

Example 7.5 Actuarial Science. Jones, aged 36 and male, wishes to buy one-year term life insurance. The cost of this insurance will be largely determined by his probability of dying, as estimated using the relative frequency for a sample of like individuals. In this very real cost-determining sense, that which is measured by relative frequency does attach to Jones as a member of a group subject to a common experimental recipe.

But Jones can be seen as being "like" different groups of individuals. If Jones purchases his insurance as an employee of his employer, then experience may indicate his probability of death to be one in a thousand. If he purchases his insurance as a member of the American Statistical Association, then the corresponding experience may be one in two thousand, and this difference will be reflected in his premium. The premium depends on the risk pool—called a cohort—to which Jones can establish himself as belonging. As an object, Jones does not have two probabilities of dying; as an object he has no (propensity) probability of dying. But probability does attach to Jones as representative of a group and if he can be seen as representing two separate groups, then the two probabilities would be expected to differ. The single instance which determines his premium refers to the experiment of picking an individual from the relevant risk pool and observing whether that individual dies. The risk pool of reference is crucial.

7.6. Locations of Some Kinds of Probabilities

All probabilities are subjective but not to the same extent. Propensity probabilities are influenced by mind but they are regarded as being about a performable experiment, and to that extent they are physical. They are ultimately unknowable but their existence in nature, in the sense of trustworthy conclusion, can be tested.

Fair betting and belief probabilities are influenced by the nature of the experiment through its effect on the judgment of the observer. For instance, odds at the track are influenced by the field a horse is racing against. But fair betting probabilities are about what an observer *thinks* about the experiment; they are personal since odds which will appear fair are a matter of personal judgment. Fair betting probabilities need have no relation to how nature is, but the correspondence can

TABLE 7.4. Locations of some kinds of probabilities

Kind	Location
Equally likely	
Laplacian	Mind of Laplacian
Physical	Experimental recipe
Fair betting (or belief)	Mind of bettor (see Section 6.2)
Frequency	
vonMises	Collective
Propensity	Experimental recipe

be checked and corrected by performing the experiment which is the object of the betting, if there is one. If there is no performable experiment then degree of belief, as usually developed in Section 6.2, does not apply.

Both numerator and denominator of equally likely probabilities-the ratio of the number of favorable outcomes to the total number of possible outcomes of an experiment-are directly influenced by the experiment. But equally likely probabilities may be either physical or personal depending on their regarded location. By adding further constructs equally likely probabilities can be seen as either propensity or subjective; but these constructs are not immediate so that equally likely probability is best thought of as a third interpretation which is sometimes the most appropriate. Table 7.4 is a useful summary.

III

Statistical Models of Induction

8

A Framework for Statistical Evidence

8.1. Introduction

We begin our discussion of statistical evidence by clarifying terms; while it won't quite serve as a definition, that aspect of statistics which we consider is concerned with models of inductive rationality. We will be concerned with rational reasoning processes which transform statements about the outcomes of particular experiments into general statements about how similar experiments may be expected to turn out. The purpose of statistical theory is to *guide* and *explain* statistical methods.

The concepts of experiment, parameter, and evidence play central roles in statistical theory and yet discussion of their meanings is often carefully avoided. What is an experiment? A parameter? What is statistical evidence about? Much statistical theory provisionally assumes the true density of data Y to be some unknown member of a family $\mathbb{F}$; but the nature of the provision is not discussed. Often, according to Basu (Ghosh, 1988, p. 12), Y is considered to be the result of an "experiment performed with a view to elicit some information about a physical quantity θ." Birnbaum (1962) adopts such a view; he writes $E_v(E, y)$ for the "evidential meaning" of obtaining data y as the outcome of experiment E. $E_v(E, y)$ is often considered to be the report to be written as a result of performing E and observing y.

The dominant framework for statistics concerns a true parameter value or state of nature.

8.1.1. The "True Value" Model

i. The raw material of a statistical analysis is an observation y on a random variable Y. The totality of possible ways in which Y might turn out, $\{y\} = S$, is called the **sample space.**

ii. the density of Y is known to be a member of the set $\mathbb{F} = \{f_\theta : \theta \varepsilon \Omega\}$ and hence can be written as f_τ where τ is the unknown **true value** of θ; τ is a fixed but unknown constant. The parameter θ is a possible candidate for τ. The set of all candidates for τ, denoted by Ω, is called the **parameter space.** Thus f_τ is the consequence of a true "state of nature" obscured by chance variation.

iii. Berger and Wolpert (1984, p. 24), and others, identify the concept of an experiment with the triple $(Y, \theta, \mathbb{F})$; they write $E = (Y, \theta, \mathbb{F})$. Thus an experiment is formulated in terms of a parameterized set of densities know to contain the true density of Y.

iv. Concerning Birnbaum's concept of evidential meaning, Berger and Wolpert (1984, p. 25) specify that $E_v(E, y)$ is the "evidence about θ arising from E and y."

In spite of its general acceptance as background for statistics—for example, Lehmann (1986) or Berger and Wolpert (1984)—we are critical of the true value model.

We first observe that the assumption (ii)—that the functional form of the density of Y is exactly known—cannot be literally correct. Hence f_τ and τ do not exist: the true density of Y will not be a member of $\mathbb{F}$ and will depend on conditions other than the state of nature, for example, the observer and his equipment. We enlarge on this in Section 8.5. Second, in the next section, we will explain that an experiment is a text, the instructions for performance, not a triple $(Y, \theta, \mathbb{F})$. As Basu says (Ghosh, 1988 p. 23), $(Y, \theta, \mathbb{F})$ is only a model, a mathematical framework, for E. Third, contrary to our usage, it is usual to not distinguish between τ and θ. The meaning of (iv) is then unclear; it probably means that $E_v(E, y)$ is about inferring which f_θ in $\mathbb{F}$ is f_τ, the true density of Y. Finally, Section 8.3 explains that a purely empirical model of induction—implied by the notation $E_v(E, y)$—does not exist; a missing ingredient is theory. Section 8.4 suggests an alternative interpretation which repairs these difficulties and 8.5 illustrates the superiority of this alternative.

8.2. What Is an Experiment?

An experiment is not a set of possible densities of a random variable. The purpose of experimentation is to ask a question of nature—conditions are imposed and nature's response is observed. Measurement is a kind of experiment; the question is, "How much of a property does a thing have?" (cf. Eisenhart, 1963). We may think of an **experiment** E abstractly as a doublet (C,Y) where C and Y describe the conditions imposed on nature and the instructions for observation, respectively. An experiment is in fact an experimental situation or setup, the recipe for performance. A performance or realization of the experiment consists of carrying out the recipe, imposing the conditions, and observing the outcome.

It is customary, as we have done in Section 5.2, to introduce the concept of probability experiment or chance situation by coin, dice, or card examples, leaving it otherwise completely undefined. This tactic conveniently sidesteps difficult issues for axiomatizing probability but is quite unsatisfactory for the purpose of providing interpretations. We need an example which is slightly more involved.

Example 8.1 Toy truck experiment. Wishing to illustrate the determination of the coefficient of friction of a toy truck, for illustrative purposes, H.E. White (1958) arranged the experimental setup of Figure 8.1 on a tabletop. A stopwatch was

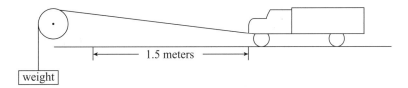

FIGURE 8.1. Coefficient of friction of a toy truck.

TABLE 8.1. Five determinations of a coefficient of friction

Observed force (F)	Observed time (t)	Calculated coefficient of friction (μ)
.196	6.05	.0016
.392	4.19	.0026
.588	3.40	.0035
.784	3.03	.0067
.980	2.54	.0026

used to observe the time required for the truck, having a mass of 2 kg, to be pulled 1.5 meters across a piece of glass. The equations $F - mg\mu = m\,a$ and $a = 2s/t^2$ are to be used to calculate μ, the coefficient of friction. F (Newton's) is the force pulling the truck, and t (seconds) is the observed time, m (kilograms) and a (meters/sec^2) are the mass and acceleration of the truck, and s $= 1.5$ (meters) is the distance traveled. The acceleration due to gravity is denoted by g.

This is the text, the recipe, for performance (C, Y), which is the experiment. Note that, to define the experiment, nothing need be said about a class of possible densities or a triple $(Y, \theta, \mathbb{F})$.

The data of Table 8.1 were observed. From five different pieces of data, five different estimates of μ, the coefficient of friction, may be calculated. The results of these calculations are given in the last column of Table 8.1. Table 8.1 is a record of five different performances of an experiment. Further theory is necessary to formulate and determine "the" coefficient of friction.

8.3. $E_v(E, y)$ Does Not Exist

Statisticians are used to thinking that they apply their logic to models of the world; less common is the realization that their logic itself is only a model.

Writing a report of the meaning of observing data y as the outcome of experiment E, in the absence of all theory, would be tantamount to a purely empirical scientific method. But it is widely accepted—Chalmers (1982, Ch. 3) and Savage (1962, pp. 14 and 15)—that theory and experiment are both essential to scientific process. Savage makes this observation to argue for prior distributions, but another part of evidential meaning is the criteria or tradition according to which the experiment is to be judged as a solution to the puzzle at hand.

The existence of $E_v(E, y)$ would imply a unique inductive logic for transforming particular to general statements, but even deductive logic—which transforms general statements into other general statements—is not unique. Till quite recently, it was believed that Aristotelian logic was the only possible deductive logic, but it is now realized—Eves (1990, pp. 166, 243)—that there are an unlimited number of possibilities and that a mathematical theory results from the interplay of two factors—a set of postulates and a logic. It isn't reasonable to expect greater specificity for induction than for deduction.

Rather than $E_v(E, y)$ we should ask about $E_v(E, T, y)$, the evidential meaning of observing y as the outcome of experiment E in the context of some theory T. An important part of T is the inductive logic being employed. Examples of inductive logics are Bayesian statistics and $p-$values. This makes it clear that $E_v(E, T_1, y)$ need not equal $E_v(E, T_2, y)$, in agreement with our Section 2.3. The true parameter model is an example of the Euclidian misconception; it assumes that conclusions from data are about evidence rather than models of evidence. Perhaps the belief is, that this error makes no difference, and hence is not worth exploring. But our further discussion suggests a change in perspective which has consequences.

8.4. The Fitted Parameter Model

A quote of Karl Pearson—found in Inman (1994, p. 6)—emphasizes the fundamental nature of the point we are making:

The 'laws of Nature' are only constructs of our minds; none of them can be asserted to be true or to be false, they are good in so far as they give good fits to our observations of Nature, and are liable to be replaced by a better 'fit'. . .

The "true value" is a philosophical abstraction—traceable to the ancient Greeks—to which we have become accustomed; but it is not verifiable since observations of it will employ specific instruments and persons. More recently, Davies and Kovack (2001, p. 3) make a similar point:

We take the point of view that models are approximations to data. In particular we make no reference to "true" regression functions for real data as we do not think that these exist in the sense that, say, elephants exist. By formulating the problem in terms of approximation we avoid the embarrassment of using the word "true."

To a statistician, Pearson's observation suggests the fitting of a density to data but there is a model building issue which is more important. The experimenters will not wish to fit just any experiment, they will be varying the text (C,Y) to achieve a condition- outcome correspondence which is "interesting," usually one which fits background knowledge or preconceived opinion. Parameters are constructs to fit an intended (target) experiment.

Further, although we prefer to think that we are measuring or discovering the "true value"—because like Everest it is there—it is partially negotiated.

In the final analysis, the "true value" of the magnitude of a quantity is defined by agreement among experts on an **exemplar method** for the measurement of its magnitude.

Eisenhart (1963, p. 30).

This negotiation is the mission of, for example, the 130 technical committees of the American Society for Testing and Materials (ASTM) that establish standards for materials, products, systems and services. More than 10,000 standard test methods, specifications, classifications, definitions, and recommended practices now in use appear in the *Annual Book of ASTM Standards* (1999). Though they might talk otherwise; ASTM committees are not engaged in discovering "true values." They are concerned with constructing experiments which are reliable in that one can predict the results of future experiments on the basis of past experiments.

If f_τ and τ do not exist, how then might we view the background of statistical inference? We suggest the alternative.

Fitted Parameter Model

i. "A conclusion is a statement which is to be accepted as applicable to the conditions of an experiment or observation...." (Tukey, 1960, p. 425). The basic concept of a statistical investigation is that of experiment, not true value.

ii. An experiment is a recipe $E = (C,Y)$ for performance, not a triple.

iii. The triple $(Y, \theta, \mathbb{F})$ is not a characterization of E, but only an assumption about E; it summarizes the background working hypothesis that observations on $E = (C,Y)$ can be fitted to some density in the class $\mathbb{F} = \{f_\theta : \theta \in \Omega\}$. The parameter θ is an index of densities being considered as candidates for fitting E, and $\mathbb{F}$ designates all densities currently being considered.

iv. The best-fitting density (BFD) of (iii) is sometimes interpreted as the true density of Y, but it only needs to be sufficiently precise for the practical purpose at hand. The BFD is not the consequence of a true state of nature obscured by chance variation. In fact it is the other way round: the BFD is a construct to fit E; hence, it is not a constant of nature but varies with E, $\mathbb{F}$ and the criterion for fitting.

v. $E_v(E, T, y)$ is not about the labeling of densities or the true value of a parameter; it is a statement—based on inference theory T—about fitting densities of the class $\mathbb{F}$ to experiment E. The assumption $(Y, \theta, \mathbb{F})$ is a part of T.

We are really concerned with fitted rather than true parameters. *The reality of the situation (the only observable truth) is the pattern of values exhibited when an experimental recipe is applied, and properties of things—fitted parameters—are constructs to describe this pattern.*

8.5. Interlaboratory Experimentation

There has been work on the nature of chance situations, particularly by Shewhart, (1939), Youden (1962), Eisenhart (1963), and Deming (1986). Their work on interlaboratory measurement makes the concept of experiment particularly clear.

Shewhart formalizes the measurement experiment as consisting of a "text." Eisenhart calls this text a method of measurement and distinguishes it from a measurement process:

Specification of the apparatus and auxiliary equipment to be used, the operations to be performed, the sequence in which they are to be executed and the conditions under which they are respectively to be carried out—these instructions collectively serve to define a method of measurement. A measurement process is the realization of a method of measurement in terms of particular apparatus and equipment of the prescribed kinds, particular conditions that at best only approximate the conditions prescribed, and particular persons as operators and observers.

Eisenhart (1963, p. 21)

Concept of a "Repetition" of a Measurement
As a very minimum a "repetition" of a measurement by the same measurement process should "leave the door open" to, and in no way inhibit changes of the sort that would occur if, on termination of a given series of measurements, the data sheets were stolen and the experimenter were to repeat the series as closely as possible with the same apparatus and auxiliary equipment following the same instructions. In contrast, a "repetition" by the same method of measurement should permit and in no way inhibit the natural occurrence of such changes as will occur if the experimenter were to mail to a friend complete details of the apparatus, auxiliary equipment, and experimental procedure employed—i.e., the written text specification that defines the "method of measurement" concerned—and the friend, using apparatus and auxiliary equipment of the same kind, and following the procedural instructions received to the best of his ability, were then after a little practice, to attempt a repetition of the measurement of the same quantity. Such are the extremes, but there is a "gray region" between in which there is not to be found a sharp line of demarcation between the "areas" corresponding to repetition" by the same measurement process, and to "repetition" by the same method of measurement.

Eisenhart (1963, p. 41)

The distinction between method and process will remain important for experimentation in general. On the one hand there is Shewhart's "text" or Eisenhart's "method" or what we have called the "recipe" (C,Y). On the other there is the process of experimentation (C,Y,j) which is a realization of (C,Y) under particular circumstances j—which we call the "laboratory." Different circumstances will be important for different experiments but will always include fixed equipment and operators. Often an important consideration is that a series of experiments is carried out with fixed "setup", calibration or orientation. Sometimes the day or time period on which the experiments are carried out is relevant. Two experimental processes following the same method will differ because of unavoidable imprecision in the text and the impossibility of carrying out precise instructions in practice; see Eisenhart (1963, pp. 165–6).

Example 8.2 The astronomical unit. Youden (1962) presents 15 determinations of the astronomical unit—the average distance between the earth and sun—along with each experimenter's estimate of spread (see our Table 8.2). (Youden does not specify the meaning of "estimate of spread.") He comments, "the best value reported by a later worker is often far outside the limits assigned by an earlier

TABLE 8.2. *Different values reported for the astronomical unit* (Values 1–12, from *Scientific American,* April 1961)

Number	Source of measurement and date	A.U. in millions of miles	Experimenter's estimate of spread
1	Newcomb, 1895	93.28	93.20–93.35
2	Hinks, 1901	92.83	92.79–92.87
3	Noteboom, 1921	92.91	92.90–92.92
4	Spencer Jones, 1928	92.87	92.82–92.91
5	Spencer Jones, 1931	93.00	92.99–93.01
6	Witt, 1933	92.91	92.90–92.92
7	Adams, 1941	92.84	92.77–92.92
8	Brower, 1950	92.977	92.945–93.008
9	Rabe, 1950	92.9148	92.9107–92.9190
10	Millstone Hill, 1958	92.874	92.873–92.875
11	Jodrell Bank, 1959	92.876	92.871–92.882
12	S. T. L., 1960	92.9251	92.9166–92.9335
13	Jodrell Bank, 1961	92.960	92.958–92.962
14	Cal. Tech., 1961	92.956	92.955–92.957
15	Soviets, 1961	92.813	92.810–92.816

*This is Table 16 of Youden (1962, p. 94).

worker." In fact, only 22 pairs of estimated intervals overlap, out of a possible 105. It seems clear that the various workers are actually estimating different quantities. The value of the astronomical unit depends on how the averaging is done: Who does the averaging? What equipment is used to measure it? What auxiliary theory is used? Is the equipment reoriented for each measurement or is a single "setup" used for all measurements? How are the measurements to be distributed over time? Does the effect of weather on measurement need to be accounted for? etc.

"The" astronomical unit depends as well on the method and process of measurement, (C,Y) and (C,Y,j), respectively. Further examples illustrating the unknown-able nature of the "true value" are provided by Deming (1986, p. 280), Shewhart (1939, pp. 66–70) and Frosch (2001, p. 8).

Deming (1986) declares that the true value does not exist. At least we must agree that it is not an operational concept. There is no way of deciding its existence, nor its numerical value if it does exist, On p. 280 he says,

… the process average will depend on the method of sampling lots, as well as on the method of test and the criteria imposed. Change the method of sampling or the method of test and you will get a new count of defectives in a lot, and a new process average. There is thus no true value for the process average. It comes as astonishment to most people that there is no true value for the speed of light. The result obtained for the speed of light depends on the method used by the experimenter (microwave, interferometer, geodimeter, molecular spectra).

And on p. 281 he further explains that there is no true number of inhabitants in a census since the number obtained will depend on the method of collection.

This observation—the statistician's dilemma—was illustrated by the 2000 United States national election. Who "really" won quickly degenerated into a morass procedure concerning recounts, ballot deficiencies, dangling chads, absentee ballot deadlines, qualified voters, and legal decisions see Frosch, 2001, p. 8). It is a political maximum that an election isn't over till the votes are counted. The declared winner depends, in part, on the manner of counting the votes, the experiment performed. We sometimes hear analysis of who "really" won; these depend on the experiment intended, the target experiment, which may differ from that performed. Different interested parties will naturally have different intentions and hence, in close elections, different conclusions about who "really" won. This disagreement need not involve dishonesty, merely different views about which votes "should" be counted.

Shewhart (1939, pp. 66–70) examines scientists' measurements of three of the fundamental constants of physical science, namely, the velocity of light c, the gravitational constant G, and Planck's constant h. He concludes, "...here...we have a sample of measurement among the most elite of pure science that do not seem to behave like drawings from a bowl of chips"; he attributes the discrepancy to "constant error."

Now, from a theoretical point of view, consider that several laboratories are available to measure a property of the same "thing." A common true value parameterization of the processes $(C, Y; j)$ is : $Y_{ij} = \tau + \beta_j + W_{ij}$, where Y_{ij} is the ith measurement carried out by laboratory j, τ is the true value of the magnitude of the property, β_j is systematic bias, and the $W'_{ij}s$, assumed independent, are observational errors. We may write σ_j^2 for the variance of W_{ij}. This model is wishful thinking; there is no knowable true τ.

For simplicity, consider only two laboratories. The normal equations, from which weighted least square estimates are calculated, reduce to $\bar{Y}_j = \tau + \beta_j$, $j = 1, 2$, where $\bar{Y}_j$ is the mean of measurements taken by laboratory j. These equations give us no hint about the value of τ, they have infinitely many solutions: $\tau = c$ and $\beta_j = \bar{X}_j - c$; $j = 1,2$. In the language of general linear hypothesis theory (Scheffe', 1959), τ is not estimable: there is no linear unbiased estimate of it. It is common to call $(\bar{Y}_1 + \bar{Y}_2)/2$, an estimate of τ but this statistic actually estimates $\tau + (\beta_1 + \beta_2)/2$, and if τ were really a property of nature, the systematic measuring error of the two different laboratories would be unrelated and usually would not sum to zero.

On the other hand, consider the fitted value parameterization of the processes (C,Y;j)

$$Y_{ij} = \mu_j + Z_{ij}$$

where $\mu_j = EY_{ij}$ and hence $Z_{ij} = Y_{ij} - EY_{ij}$ is experimental error. The expectations and experimental errors of the two laboratories may be anticipated to differ. Now $Y_{ij} = \mu + \delta_j + Z_{ij}$ where $\mu = (\mu_1 + \mu_2)/2$ and $\delta_j = \mu_j - \mu$. This parameterization assumes only that the expectation or some other measure of central tendency exists. The normal equations now reduce to $\bar{Y}_j = \mu + \delta_j$; $j = 1, 2$.

Note that $\delta_1 + \delta_2 = 0$ so that $(\bar{Y}_1 + \bar{Y}_2)/2$ estimates μ; but μ is not τ, the true value of the property being measured, rather it is a compromise being fitted to the systematically different determinations of the two laboratories.

The above compromise is appropriate to a target population of two laboratories available to perform the experimental method (C,Y) equally often. A reasonable model, where the recipe does not specify which laboratory will be used, is that the laboratory is chosen at random. Then, the expectation of the randomly chosen measurement is $(\mu_1 + \mu_2)/2$. The parameter of the method (C,Y) is $\mu = (\mu_1 + u_2)/2$.

In summary, interlaboratory measurement illustrates that, as Karl Pearson and others state, statistics is not about inferring true values τ; they are unknowable (even unidentifiable.) Instead it is about μ, a data fitting property of an experiment (method and or process) on nature: f_μ must fit sufficiently well for the practical purpose at hand and at least as well as any other f_θ currently under consideration. We might then call μ the best-fitting value. Experiments are not defined by parameters. It is the other way round; parameters are defined in terms of models of experiments.

8.6. Conclusion

The current framework for statistics—the true value model—has several nonintuitive features. These discrepancies have been pointed out by various authors but have attracted little interest. Perhaps the belief is that they make little difference and hence are not worth exploring. We offer an alternative framework which repairs these discrepancies. We may think either that the true value does not exist or that it depends on the target experiment and hence is not "the state of nature." This may just be calling things by their correct names but the alternative interpretation has consequences for both statistical theory and practice.

9

A Critique of Bayesian Inference

9.1. Randomness Needs Explaining

Rubin (1984) describes statistical inference to be Bayesian, if known as well as unknown quantities are treated as random variables—knowns having been observed but unknowns unobserved—and conclusions are drawn about unknowns by calculating their conditional distribution given knowns from a specified joint distribution.

A more formal version of Rubin's Bayesian inference in the discrete case is as follows: Two quantities, Y and θ, with ranges S and Ω, respectively, are treated as random variables with specific joint probability p(y, θ). Y = y is observed and therefore known; θ is an unknown quantity about which we wish to make an inference. The inferential conclusion to be drawn is that the **posterior** probability, the conditional probability of θ given Y = y, is $\pi(\theta \mid y) = p(y, \theta)/p(y)$, where $p(y) = \sum_{\theta \in \Omega} p(y, \theta)$.

For continuous random variables (r.v.s.) a similar development yields the same inferential conclusion except that probabilities are replaced by density functions.

The usual way of specifying p(y, θ) is by specifying the prior and the likelihood. The **prior,** $\pi(\theta)$, is the marginal probability of θ. The sampling density is the conditional probability of Y given θ, it is specified to be some member of a parametric class $\mathbb{F} = \{f(y \mid \theta) : \theta \in \Omega\}$. The **likelihood** is the sampling density considered as a function of θ. The joint probability $p(y, \theta)$ is specified since it is the product of the specified sampling density and prior. Now, our inferential conclusion becomes

$$\pi(\theta \mid y) = f(y, \theta)\pi(\theta)/p(y);$$

the posterior probability is proportional to the product of prior and likelihood.

Some history provides perspective. Bayes' own illustration was published in 1764–65. A modern version, based on Todhunter (1949, p. 294), first published in 1865, follows: Denote by AB one side of a rectangular billiard table, the opposite side being one unit away. Let X ($0 \leq X \leq 1$) be the distance from AB of a ball randomly thrown on the table. In n further independent random throws of the ball on the table, let S denote the number of throws on which the ball is closer to AB than on the original throw. For $0 \le b \le c \le 1$,

$$\text{pr}(b \le X \le c, S = s) = \int_b^c \text{pr}(S = s | X = x)\, dx = \int_b^c \binom{n}{s} x^s (1 - x)^{n-s}\, dx,$$

and in particular

$$\text{pr}(S = s) = \int_0^1 \binom{n}{s} x^s (1 - x)^{n-s}\, dx.$$

Combining the above two equations we obtain

$$\text{pr}(b \le X \le c | S = s) = \frac{\displaystyle\int_b^c \binom{n}{s} x^s (1 - x)^{n-s}\, dx}{\displaystyle\int_0^1 \binom{n}{s} x^s (1 - x)^{n-s}\, dx}.$$

This, strictly speaking, is Bayes' theorem; it is a continuous version of the finite result which we have given that name in Section 5.2.

About 1774 Laplace initiated the use Bayes' ideas to estimate the probabilities of causes from observed events. One of Laplace's results, called his rule of succession, is that if an event has happened n times without failure, then the probability that it will happen on the next trial is $(n + 1)/(n + 2)$. Laplace has illustrated the use of his rule by calculating the probability that the sun will rise tomorrow given that it has risen each morning for a recorded history of 5,000 years or 1,826,213 days. He obtains odds of 1,826,213 to 1 for the required event. Laplace's rule can be derived from the assumption of Bernoulli trials with uniformly distributed success probability. Write θ for success probability, $X_i = 1$ or 0 to indicate success or failure on trial i and $Y_n = \sum_{i=1}^n X_i$. Then

$$pr(X_{n+1} = 1, Y_n = n) = \int_0^1 \theta \cdot \theta^n d\theta = (n + 2)^{-1},$$

$$pr(Y_n = n) = \int_0^1 \theta^n d\theta = (n + 1)^{-1},$$

and

$$pr(X_{n+1} = 1 \,|\, Y_n = n) = (n + 1)/(n + 2)$$

There are several approaches to Bayesian inference, depending on what is meant by "treating" a quantity as a random variable. First, there is unexplained Bayesianism where, as a matter of faith, one simply *takes* *all* quantities to be random variables. But, as Deming points out (our Section 7.1) that chance is not a necessary consequence of indeterminism. Consideration needs to be given to whether Kolmogorov's axioms are satisfied.

Todhunter (1949) criticizes unexplained Bayesianism. Referring to Bayes' original problem, he says,

It must be observed ... that ... we know that a priori (any value of x between zero and one) is equally likely; or at least we know what amount of assumption is involved in this supposition.

He then contrasts this with the rule of succession:

In the applications which have been made of Bayes' theorem, and of such results as that which we have taken from Laplace ... there has however often been no adequate ground for such knowledge or assumption.

Unexplained Bayesianism is immediately suspect since it is a theorem without hypotheses—a procedure to be followed without explanation. Later, in Table 9.2, we provide a counterexample to unexplained Bayesianism: an instance where an initially promising quantity does not conform to the probability axioms. Other approaches to Bayesian inference attempt to explain why we may treat the relevant quantities as random variables. We like this kind of Bayesian theory; if a prior distribution can be motivated, then by all means use it.

Rubin's description uses the probability concepts of random variable and probability distribution. Such entities have no extra mathematical meaning or consequence until they are interpreted. This is not a criticism of Rubin's description; it is the nature of all mathematics (Eves, 1990, p. 149). While the major interpretations of probability are in terms of chance experiments, probability need not have anything to do with chance; areas of subsets of the unit square satisfy the axioms. But if Bayesian inference is to be considered an *explanation* of why specific conclusions follow from specific data then it must conform to *some* interpretation of probability, for Bayes' result is a theorem of probability. We take the position that some quantities are reasonably modeled as following the rules of probability and some are not, and the onus is on the declarer- the user of Bayesian statistics—to provide some explanation of why a quantity may reasonably be considered random.

9.2. How to Adjust Your Belief

There are several explanations of Bayesian inference. First, we may employ the Chapter 6 interpretation of probability as personal degree of belief through consideration of "economic man." Our subsequent discussion provides necessary and sufficient conditions that economics does explain Bayesian inference. Any such analysis must involve economic ideas but we strive for an elementary and self-contained treatment which assumes no prior knowledge of economic theory.

A nice economic argument—Chapter 6 is our version—indicates that a person's risk-neutral pricing system on various wagers concerning Y and θ is a probability, say, p(y,θ). Savage (1962, Sect. 2) is an appropriate early reference; he presents the case for subjective probability as follows: "Roughly speaking, it can be shown that ... a probability structure $P_r \cdots$ exists for every person who behaves coherently in that he is not prepared to make a combination of bets that is sure to lose; the structure is such that

$$P_r(A)/P_r(\text{not } A) = P_r(A)/\{1 - P_r(A)\}$$

is the odds that he would barely be willing to offer for A against not A. "The concept of (equilibrium) price cannot be altogether escaped by anyone who would think of his own or other people's economic behavior." Therefore, since "'opinion,' when analyzed, is coterminal with 'odds,'" degree of belief is to be quantified as probability (see Chapter 6 for details). By assumption A4 of Section 6.2, p(y, θ) quantifies personal degree of belief in the joint outcome $Y = y$, $\Theta = \theta$. Therefore, by the rules of probability, degree of belief in $\Theta = \theta$ changes from $\sum_{y \in S} p(y, \theta)$— the **prior** density—before data, to the conditional density $p(y, \theta) / \sum_{\theta \in \Omega} p(y, \theta)$— called the **posterior**—upon observing $Y = y$.

As Rubin prescribes, the known y and the unknown Θ are both treated as random variables—y having been observed but Θ unobserved—and conclusions about Θ are drawn by calculating the conditional distribution of Θ from the joint density p(y, θ).

An individual's posterior degree of belief in the hypothesis $\Theta = \theta_0$ relative to $\Theta = \theta_1$ is determined by the ratio

$$\frac{\pi(\theta_0 | y)}{\pi(\theta_1 | y)} = \frac{f(y | \theta_0) \, \pi(\theta_0)}{f(y | \theta_1) \, \pi(\theta_1)}. \tag{9.1}$$

The individual will tend to believe θ_0 over θ_1 if this ratio exceeds one, and relative strength of belief increases with the ratio; the ratio of posterior probabilities quantifies the Bayesian evidence—grounds for belief—of y for θ_0 over θ_1.

In summary, the coherence argument—put your money where your (risk-free) belief is—leads to the Bayesian result that belief is to be updated, to reflect new data, according to Bayes' theorem.

Example 9.1 Prisoner's dilemma (continued). The discussion of Example 7.1 satisfies Ruben's specification: Writing Θ and R for prisoner to live and jailer's response, then R is known but Θ is unknown and both are to be treated as random— $R = r$ having been observed but Θ still unobserved. Conclusions about Θ are to be drawn from the conditional distribution of Θ given $R = r$.

The subjective degree of belief interpretation is that, before the jailor's response, A believes the three possible outcomes of Θ are equally likely; hence the prior probability that A lives is 1/3. The probability of response B given A lives is p, as calculated from Table 7.2. From Bayes rule, A's posterior degree of belief that he will live is p/(1 + p) as before.

9.3. The Economic Approach to Group Belief

Fair betting price or probability may be interpreted as individual degree of belief, but whose utility and whose belief? The statistical analyst's? His client's? Or perhaps the belief of an entire scientific community? That scientific belief is more social than personal is a major twentieth-century finding of the philosophy of science; see our Chapter 3 where we have discussed the work of Kuhn (1962), Toulmin (1972), and Hull (1990). A personal Bayesian finding is open to the comment: "You are welcome to your opinion, but what is your belief to me?" For some purposes what is important is not so much individual belief, but the shared beliefs of a group.

Other literature on the economic approach to coherence follows two main themes. First, Nau and McCardle (1991) develop a complete descriptive theory of economic decision-making, not limited to risk neutral agents. They take the view that the separation of belief from risk preference "is inessential to the characterization of economic rationality in terms of observable behavior." Their paper is about behavior; they sidestep the concept of belief and therefore say nothing about how belief can be quantified.

The second theme develops and studies the properties of what Genest and Zidek (1986) call the "supra-Bayesian approach." A good explanation of the supra-Bayesian position can be found in D.V. Lindley (1985). The bulk of his book is about coherent individual decision-making. Then in a closing chapter, Lindley argues eloquently that providing a theory of group decision-making is an important unsolved problem. He continues, on page 180—

...notice that it is possible to think of the committee as a decision-maker and for it to play the role of someone desirous of viewing the world coherently; of not violating the sure-thing principle; and, hence, of producing probabilities and utilities of its own. There is nothing in this book that requires the decision-maker to be a person: the theory concerns the choice of an action and the chooser could be a committee. There is just as much reason for a committee to be rational as for an individual. In other words, our problem can be thought of as passing from a set of values, one for each committee member, to a single valuation. How is this to be done? At present nobody knows.

Many authors use the word *pooling* to describe the process of passing from a set of values to a single valuation. The purpose of the pooling may be either to achieve consensus or to summarize. Achieving consensus implies changing individual opinions to a common view. A summary leaves individual opinions unchanged but restates them more succinctly while retaining their overall sense. DeGroot (1974, p. 119) advances an intuitive Markov chain model for "reaching a consensus." His consensus takes the form of a subjective distribution function,

$$\sum_{i=1}^{k} \Pi_i F_i$$

where F_i is the pre-consensus subjective distribution function of the ith of k individuals. Since the Π's are positive and sum to one, DeGroot's consensus is a subjective distribution common to all individuals. DeGroot's Markov assumption is a clever and intuitive sociological model, but individuals might not reach consensus by this or any mechanism, nor is it clear that they should. Our discussion is about summarization.

Weirahandi and Zidek (1981) describe the group coherence program as aggregation; individual "assessments are combined in some way into one that may be used in a conventional uni-Bayesian analysis." The supra-Bayesian approach therefore assumes that group belief can be quantified as a probability. But G.A. Barnard (1980) is skeptical:

> The modified personalistic view of probability put forward by Box and Borel is, I think, inadequate for statisticians basically because statisticians work for clients. They are therefore not concerned with personal probabilities but with what might be called 'agreed probabilities.' And whereas personal probabilities may be said always to exist in principle, agreed probabilities need not exist.

Roberts (1965) and McConway (1981) show us where to look for Barnard's "agreed probabilities." Define a rule for pooling or summarizing individual probabilities to have the strong setwise function property (SSFP) if there exists a function f such that for each event A, if $a_1, \ldots a_l$ are the individual probabilities then $f(a_1, \ldots a_l)$ is their summary. For finite sample spaces this is equivalent to McConway's (1981, p. 412) definition. Restriction to SSFP pooling is intuitive in the present context since it says that summarized group degree of belief about an event depends only on the individual members' degrees of belief about the event.

An example of a pooling rule having the SSFP is the linear opinion pool, $f(a_1, \ldots a_l) = \Sigma_j w_j a_j$, where the weights are nonnegative real numbers summing to one. Roberts (1965) shows that if individual unconditional probabilities are summarized by a linear opinion pool, and the combined summary is to satisfy the probability calculus, then the group summary of $P(B|A)$ must be

$$P(B|A) = \sum_j w_j a_j c_j / \sum_j w_j a_j \tag{9.2}$$

where a_j and c_j are $P(A)$ and $P(B|A)$ for the jth individual. Conditional probabilities are summarized linearly but the weights are revised and depend on individual unconditional probabilities. With this introduction we may state the following result.

Theorem 9.1 For a sample space containing at least three points, if the unconditional probabilities of a group summary satisfy the SSFP then the completed summary will be a probability if and only if the summary of unconditional probabilities is a linear opinion pool and the summary of conditional probabilities follows Roberts' rule for weight revision.

Proof of necessity
McConway's (1981) Theorem 3.3 implies that, for any SSFP pooling rule, the group summary of $P(A)$ will be a linear opinion pool. This places us in Roberts' situation where, to satisfy the multiplication rule, we must have equation (9.2).

Theorem 9.1 tells us that probabilities do exist, on which the members of a group may agree. But to adopt that expression of belief without some justification leaves Bayesian inference unexplained. Unfortunately, there is a problem with representing group belief as probability. The usual (Chapter 6) coherence argument for persons, already has implications for groups. The reason is that gambling is a group activity; an individual cannot place a bet unless someone else "covers" his bet. Price—which is exchange rate—is a group concept; we cannot contemplate the purchase of A or Ā tickets unless individuals have wealth and a market for exchange exists, at least conceptually. A single individual cannot engage in exchange. A system of markets and prices is implicit in the price interpretation of probability.

In one sense, Barnard is correct; the economic approach to degree of belief is generally incompatible with supra-Bayesianism. For example, Seidenfeld et al. (1989) show that two Bayesians with different probabilities and utilities have no Bayesian Pareto compromises except to adopt one or the other's approach entirely. Perhaps this settles the matter. The economic explanation of Bayesian statistics visualized by Savage and Lindley doesn't work in general. But, for completeness, two further questions are of interest: (i) when *does* the economic approach lead to Bayesian inference? and (ii) where *does* the economic approach lead for groups?

To apply equilibrium price to betting, as Savage suggests, we may view gambling on the outcome of a chance experiment E as a transaction in which two gamblers purchase amount $c + d$ of A and Ā tickets at agreed costs of c and d. The significance of an amount q of A tickets is that a pot of amount q is set aside until E is performed; the bearer receives the entire pot or nothing according as A does or does not occur.

Consider a market for A and Ā tickets which is subject to Savage's probability structure. Each individual bettor evaluates his bets solely by comparing market price which he considers fixed or outside of his control with his personal unit price—determined from his personal odds. He does not bluff, threaten, bargain, etc. This is the definition of a competitive market. The group will prefer to behave as a competitive market. This is a consequence of the personal probability argument, not of economic theory.

We will be concerned with the holdings—of the form (x,y) where x and y are amounts of A and Ā tickets—of a gambler who is endowed with amount m of money. Denote by $a(\bar{a} = 1 - a)$ the gambler's personal probability of $A(\bar{A})$ and let $p(\bar{p} = 1 - p)$, $0 < p < 1$, signify the prevailing $A(\bar{A})$ ticket price. If $0 < p < a$, the gambler will wish to buy no Ā tickets and as many A tickets as possible; the situation is analogous for $a < p < 1$. However, if $p = a$, then any portfolio of the form $(tm/p, \bar{t}m/\bar{p})$ will optimize his preference (t is the proportion of the total tickets purchased that are A tickets and $\bar{t} = 1 - t$). Table 9.1 is a summary.

TABLE 9.1. Individual demands for A and Ā tickets

Condition	t	x	y
$0 < p < a$	1	m/p	0
$p = a$	$0 \leq t \leq 1$	tm/p	$\bar{t}m/\bar{p}$
$a < p < 1$	0	0	$m/\bar{p}$

Group demand is determined by summing over all participating individuals. Thus if j designates the jth participant, $j = 1, \ldots, l$, then

$$x(p) = \left(\sum_{\{j:p<a_j\}} m_j + \sum_{\{j:p=a_j\}} t_j m_j \right) \Big/ p$$

is the quantity of A tickets that all participants would wish to hold if the price were p. Similarly,

$$y(p) = \left(\sum_{\{j:p>a_j\}} m_j + \sum_{\{j:p=a_j\}} \bar{t}_j m_j \right) \Big/ (1-p)$$

is the demand for $\bar{A}$ tickets if their price were $1 - p$. Choice of p, the ticket price faced by the group, divides the group into potential buyers of A tickets and potential sellers. Total amount exchanged will be $q_0 = \min[x(p), y(p)]$ since amount sold must equal amount bought.

Equilibrium price is achieved when the demands for A and $\bar{A}$ tickets are equal, $x(p) = y(p)$. This simplifies to

$$\sum_{\{j:p<a_j\}} m_j + \sum_{\{j:p=a_j\}} t_j m_j = p \sum_{1}^{l} m_j,$$

$$F(p) - \bar{p} = \sum_{\{j:a_j=p\}} t_j f_j \tag{9.3}$$

where $f_j = m_j/M$, $M = \sum_{1}^{l} m_j$, $F(p) = \sum_{\{j:a_j \leq p\}} f_j$:, and $F(p-) = \sum_{\{j:a_j < p\}} f_j$.

Theorem 9.2 An equilibrium price p_A exists and is the solution of the inequalities

$$F(p-) \leq \bar{p} \leq F(p). \tag{9.4}$$

Proof
Since $0 \leq t_j \leq 1$, any solution of (9.3) must satisfy (9.4). The inequalities (9.4) have a unique solution since $F(p)$ is a nondecreasing jump function with $F(0) = 0$ and $F(1) = 1$ while the function $\bar{p} = 1 - p$ decreases from 1 at $p = 0$ to 0 at $p = 1$. The solution p_A is the value of p, where $z = F(p)$ intersects the line $z = 1 - p$. p_A is an equilibrium price since we may take $t_1 = \ldots = t_l = 0$ if $F(p_A) = \bar{p}_A$, and $t_1 = \ldots = t_l = [F(p_A) - \bar{p}_A]/[F(p_A) - F(p_A-)]$ if $p_A < F(p_A)$.

The novelty of Theorem 9.2 is that (9.4) does not hold in a general economic setting, only for the very special market having Savage's structure.

Equilibrium price is not in general a probability; counterexamples are easy to construct. For the special personal probabilities of Table 9.2 the equilibrium prices—calculated from (9.4) assuming equal monetary endowments—of A and B do not sum to that of $A \cup B$. In passing, this illustrates the error of unexplained Bayesianism- uncritically assuming all quantities to be random. Savage sought to base Bayesian inference on economic behavior; but the central economic quantity, equilibrium price, cannot be "taken" to be random.

TABLE 9.2. Equilibrium price is not a probability

Events	Individual probabilities		Equilibrium price
	1	2	
A	.2	.3	.3
B	.1	.4	.4
C	.7	.3	.5
A ∪ B	.3	.7	.5

As a summary of group belief, equilibrium price has an optimal property. Let us suppose that the satisfaction of a group member with a group summary p of personal A probabilities is demonstrated by his preferred gambling action at A ticket market price p. Specifically, if he will buy, then his satisfaction is measured by the amount he was willing to pay minus the amount he does pay; similarly, the satisfaction of a seller is measured by amount received minus amount at which he would have sold.

Theorem 9.3 Equilibrium price is the group summary that maximizes group satisfaction.

Proof
If $x^{-1}(q) > p$, then some group member is willing to pay $x^{-1}(q)dq$ for amount dq of A tickets, but his cost is only pdq, $[x^{-1}(q) - p]dq$ is then a measure of that member's satisfaction with that purchase and $\int_0^{q_0} [x^{-1}(q) - p]dq$ is the total satisfaction of all buyers. Similarly, $\int_0^{q_0} [p - y^{-1}(q)]dq$ is the satisfaction of all sellers and, therefore, $S(p) = \int_0^{q_0} [x^{-1}(q) - y^{-1}(q)]dq$ is the satisfaction of all group members with the exchange that occurs at price p. The signed area $S(p)$ is maximized when $p = p_A$, as is suggested by Figure 9.1. (Actual curves are however discontinuous at individual prices.)

The novelty of Theorem 9.3 is the reinterpretation of a standard economic argument for statistical inference.

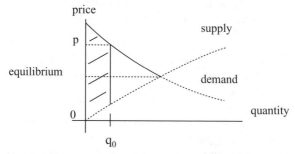

The crosshatched area represents total group satisfaction with summary value p.

FIGURE 9.1. Preferred group gambling behavior for a lottery ticket.

Barnard's criticism of economic probability comes in two parts. First, the personal perspective is inadequate for much of statistics. The statistician must satisfy his client(s) and the prospective author of a scientific article must convince a peer group of referees and editors. The attitude of one individual to the unsupported opinions of another will be "What is your belief to me?" Much of statistical practice is not about personal belief or behavior, it is about convincing, and hence involves multiple persons. Any economic analysis of statistical inference needs to extend to the beliefs of a group.

We think Savage is right about equilibrium price being the natural way to extend (economic) personal beliefs to groups. In review: The personal theory already contains implications for groups. To employ economic man to explain personal belief but reject him when considering public belief would not be coherent in the dictionary sense of approaching all problems from a single set of principles. Equilibrium price is the most important single feature concerning the preferred economic behavior of a group of individuals called a market. Equilibrium price has the optimal property of Theorem 9.3. Savage does have the problem of specifying the distribution of initial wealth, but let us suppose that problem to be solved—perhaps by the democratic criteria of equality.

A criticism is that equilibrium price cannot be a rational summary of a group belief as evidenced by economic behavior since it is not a probability and therefore a clever outsider can make a Dutch book against the group. But this ignores the constraints of the situation; there *is* something in Lindley's book which requires the decision-maker to be a person. An outsider cannot simply make a bet; he must find someone willing to take the other side. To make a bet the outsider must become a member of the competitive (according to Savage's structure) market where his proffered bet may or may not be accepted.

Whether the second part of Barnard's criticism—that agreed probabilities need not exist—is correct seems to depend on purpose. In the very important task of model building, the model assumptions provide probabilities on which everyone is to be conditionally agreed. For example, Bayes' billiards problem prescribes conditionally that the position of the initial ball is uniformly distributed. Of course, if the individuals of a group are of a consensus then the group as a whole can agree on their common view. This follows from our Theorem 9.2.

The present Chapter accepts the economic Bayesian result for persons and for conditional model building but is frankly critical of its extension to activities where diverse personal views are to be reconciled. Note, in passing, that here we have arrived at the same conclusion as the "generally accepted" requirement for legal evidence to be admissible—compare Section 4.1. Combining this with Chapter 6 findings, we conclude that the fair betting utility interpretation of probability provides an explanation of Bayesian inference when and only when the concern is a chance experiment which does eventuate and all individuals concerned are agreed on the relevant probabilities. The same considerations which indicate probability as an expression of personal belief, point toward equilibrium price for groups.

9.4. Objective Bayesian Statistics

Subjective Bayesianism has lost ground in recent years, partly because of the failure (for groups) of Savage's program to base inference on "economic man" but primarily due to a desire to be "objective." Scientists want physics to be about matter or energy rather than about the psychology or sociology of physicists.

A definition of "objective" in this context is hard to find. Is a statistical analysis "objective" just because it is not subjective and therefore represents no ones belief? And if so, what is the virtue in that? "Objective" is not the same as "nonimformative." Ni and Sun (2003, p. 160) write "Nonimformative priors are designed to reflect the notion that a researcher has only vague knowledge about the distribution of the parameters ..." and on p. 162 they refer to a nonimformative prior as an "expression of ignorance." It takes a mind to be uninformed or ignorant or to have vague knowledge. The noninformative prior is a conceptual subjective concept. We think that in striving to be "objective," scientists are looking for an analysis which arrives at conclusions about propensity probabilities, about the tendencies for events to occur when experiments are performed.

An explanation of the prior, which is subject to a frequency interpretations, is that we are fitting a sequence of performances of an experiment by a class of exchangeable r.v.s. This impressive explanation is based on deFinetti's theorem.

If the joint distribution of a finite number of the random variables $X_1, X_2, \cdots$ does not depend on which are chosen, but only on how many are chosen, then the random variables are **exchangeable.** That is, the sequence is exchangeable if for every $k_1, \cdots, k_n$ and every m, we may write $pr(X_{k_1} \leq x_1, \cdots, X_{k_m} \leq x_m) = G_m(x_1, \cdots, x_m)$, Random variables which are *independent and identically distributed (i.i.d.)* provide an example of exchangeability.

Theorem 9.4 (de Finetti) The concept of exchangeability is equivalent to that of conditional independence with common $d.f.$

An outline of a proof due to Loéve' (1955, p. 365) follows:

Proof
The empirical $d.f. F_n(x) = n^{-1} \sum_{i=1}^n I[X_i \leq x]$ approaches a random function $F(x)$ which has the properties of a distribution. For, $F_n(x).F_m(x) = (nm)^{-1} \sum_{i,i'} I[X_i \leq x, X_{i'} \leq x]$ and, for $m \leq n$, $E F_n(x) F_m(x) = \frac{1}{n} G_1(x) + \frac{n-1}{n} G_2(x, x)$; thus $E[F_n(x) - F_m(x)]^2 = \frac{n-m}{nm} [G_1(x) - G_2(x, x)] \xrightarrow[m \to \infty]{} 0$. Therefore $F(x)$ exists such that $F_n(x) \to F(x)$ in probability. Since $F_n(x)$ is a distribution function (d.f.), the limit $F(x)$ also has those properties. By the dominated consequence theorem, for $k = 1, 2, \cdots, m$, $E F_n(x_1) \cdots F_n(x_k) \xrightarrow[n \to \infty]{} E F(x_1) \cdots F(x_k)$

On the other hand $E F_n(x_1) \cdots F_n(x_k) \xrightarrow[n \to \infty]{} G_k(x_1, \cdots, x_k)$. Consider the case $k = 2$, $F_n(x_1) \cdot F_n(x_2) = (n^{-2}) \sum_{i,i'=1} I[X_i \leq x_i, X_{i'} \leq x_{i'}]$ and $E F_n(x_1) \cdot F_n(x_2) = \frac{n(n-1)}{n^2} G_2(x_1, x_2) + \frac{n}{n^2} G_1[\min(x_1, x_2)] \to G_2(x_1, x_2)$.

Similarly,

$$EF_n(x_1)F_n(x_2)F(x_3) = \frac{n^3 - n(n-1) - n}{n^3} G_3(x_1, x_2, x_3) + 0(n)$$
$$\to G_3(x_1, x_2, x_3)$$

The general result is similar but more complicated. Finally, putting the two limits together, $G_k(x_1, \cdots, x_k) = EF(x_1) \cdots F(x_k)$, where the expectation is with respect to the randomness of the $d.f.$ F.

The conditional distribution of $X_1, \cdots X_k$ given that F_n approaches F in probability is $F(x_1) \cdots F(x_k)$; the unconditional distribution is $EF(x_1) \ldots F(x_k)$. Denote the marginal probability density of $\lim F_n = F$, over the space of all distributions, by $\prod_F(D)$. In usual applications $\prod_F$ is either a discrete probability or an ordinary derivative but in general, a Radon–Nikodyn derivative is required.

Now the posterior density, the conditional density of F given $X_1 = x_1, \cdots, X_k = x_k$, is

$$\prod_F(D \mid x_1, \cdots, x_k) = \frac{d(x_1) \cdots d(x_k) \prod_F(D)}{E_F f(x_1) \cdots f(x_k)},$$

where d and f are densities of the distributions D and F, and $\prod_F$—the prior—is the marginal density of $\lim F_n = F$. The theorem of de Finetti allows us to interpret the prior as the marginal probability of the limiting empirical distribution function of a sequence of exchangeable $r.v.s$.

It is customary to assign positive probabilities only to a convenient standard parametric class $\mathbb{F} = \{D_\theta : \theta \in \Omega\}$, such as the normal or beta. This is equivalent to taking θ as a random quantity rather than F as a random distribution; it corresponds to a belief that some member of $\mathbb{F}$ will provide an adequate fit. Our Bayesian inference becomes $\prod_\Theta(\theta \mid x_1, \cdots, x_k) \propto d(x_1 \mid \theta) \cdots d(x_k \mid \theta) \prod_\Theta(\theta)$, in agreement with Rubin. This frequency explanation of the randomness of θ is however not helpful in choosing the prior. In particular, presumably a nonimformative prior is out of place in a frequency theory. How to choose an objective prior is an important problem for statistical theory.

It is sometimes said that the difference between sampling and Bayesian statistics is that no prior probability exists for the sampling theorist. But that is not the distinction. Bayesian theory is applicable to the sampling theorist's model, but it yields no inferential insight. The sampling theorists model is that $X_1, \cdots, X_n$ are $i.i.d.$ according to some fixed distribution, say, F. Hence, the empirical distribution F_n approaches F with probability one (see Loéve, 1955, p. 20). For the sampling theorist, the marginal density of $\lim_{n \to \infty} F_n$ is degenerate:

$$\prod_F(D) = \begin{cases} 0, D \neq F \\ 1, D = F \end{cases}$$

where F is some fixed but unknown distribution. Now, following de Finetti's reasoning, the conditional density of F given the data is

$$\prod_{F}(D/x_1, \cdots, x_k) = \frac{d(x_1) \cdots d(x_k) \prod_F(D)}{f(x_1) \cdots f(x_k)} = \begin{cases} 0, D \neq F \\ 1, D = F \end{cases}$$

The posterior probability of F also assigns unit probability to F; but we don't know F.

From this point of view, the appropriateness of Bayesian inference depends on one's assumptions about repeated performances of an experiment: if we believe that the successive observations are exchangeable and independent according to some fixed but unknown distribution then an unknown one point prior leads to an unknown one point posterior, a solution which yields no inferential insight. The limiting empirical distribution of a series of exchangeable $r.v.s$ will be probabilistically distributed over some nondegenerate space of distributions only if the $r.v.s$ are dependent. Only in this case will Bayesian inference for exchangeable $r.v.s$ be insightful.

10

The Long-Run Consequence
of Behavior

10.1. Decision Theory

As a consequence of Wald's (1950) powerful work, statistics was for a time defined as the art and science of making decisions in the face of uncertainty. The decision problem assumes the questionable true value model of Section 8.1 and contemplates deciding, on the basis of data, between various possible actions when the state of nature is unknown. It is anticipated that the data will be helpful in choosing an action since the probabilities of data depend on the state. The task of the decision maker is to choose a decision rule specifying the action d(y) to be taken if data y is observed. The theory assumes known numerical losses l (a, θ) in taking each action a when each state θ "obtains."

If τ, the true state of nature, were known then it would be reasonable to take that action which minimizes the expected loss for that state; that is, choose a decision rule d(y) which minimizes the risk, or expected loss

$$r(\tau, d) = \sum_{y \in S} l[d(y), \tau] f_\tau(y)$$

for the known state, τ. Then, as a consequence of the law of large numbers, if $f_\tau(y)$ can be given an objective interpretation, minimizing r(τ,d) can be viewed as an attempt to adopt single instance behavior which will yield small total loss for a long sequence of actual future decision problems. This behavioral criterion is one concept of what constitutes a good statistical procedure.

But the essence of the decision problem is that we do not know the true state and therefore must compromise, choosing an action which minimizes all risks, r (θ,d), $\theta \in \Omega$, "on balance." At this point, theory and practice divide into criteria and cases.

If degree of belief (Section 9.3) probabilities, π (θ), are available for the state of nature then it would be reasonable to maximize believed value by minimizing the expected risk

$$\sum_{\theta \in \Omega} r(\theta, d) \pi(\theta).$$

In the next two Sections we examine two common behavioral alternatives to this Bayesian view.

10.2. The Accept–Reject Theory

We present only the barest outline of hypothesis testing, the dominant theory of statistics. Our reasons are that the topic is extensively treated elsewhere, the reader will already have some familiarity with hypothesis testing, and our story is primarily about other theories. We discuss the topic for completeness and later comparison.

The recognized source for the mathematics of hypothesis testing is Lehmann (1986). But Lehmann does not tell us how the theory is to be applied or interpreted. It is made clear, on Lehmann's pages 68 and 69, that the true value model $(Y, \theta, \mathbb{F})$ is known to hold and that hypothesis testing is for the purpose of deciding, in the sense of Wald, whether to "accept" or "reject" an hypothesis; $\tau \in H$ where $H \subset \Omega$. But we are not told which meanings of accept and reject are intended.

With any reasonable interpretation, if H is true (false) then accepting (rejecting) H is the correct action to take and rejecting (accepting) H is an error; this is nicely summarized in the familiar Table 10.1. So accept (reject) H means "act like H is true (false)."

Table 10.1 suggests the losses of Table 10.2.

The risks are then

$$r(\tau, d) = \begin{cases} a \; P_\tau(\text{accept } H), \tau \in \Omega - H \\ b \; P_\tau(\text{reject } H), \tau \in H. \end{cases}$$

But the real consequences of actions literally interpreted will rarely have the simple form of Table 10.2. Perhaps an inferential interpretation is more natural. Lehmann (1986, p. 4) suggests, "formally [an inference] can still be considered a decision problem if the inferential statement itself is interpreted as the decision to be made." But details of the actions and their consequences are not immediate. We might take

TABLE 10.1. Errors of the first and second kind

Action	State of nature	
	H	Ω-H
Accept H	Correct	Type II
Reject H	Type I	Correct

TABLE 10.2. Hypothesis-testing losses

Action	State of nature	
	H	Ω-H
Accept H	0	a
Reject H	b	0

"accept" and "reject" to mean "believe true" and "believe false" respectively. And we might then consider l not as a literal loss but simply as an indicator of whether an inference is correct or not; summations of indicators over a series of inferences would then be the total number of incorrect inferences and minimizing the risk, in some sense, would be for the purpose of controlling the long run proportion of errors of the two kinds.

Choosing a decision procedure by minimizing the risk is still not a well-posed mathematical problem since test properties will depend on the unknown parameter τ. The hypothesis testing strategy for dealing with this basic difficulty is to adopt the following:

Strategy

Seek a test which minimizes

$$\sup_{\theta \in \Omega - H} P_\theta(\text{accept } H) \tag{10.1}$$

subject to the constraint

$$\sup_{\theta \in H} P_\theta(\text{reject } H) \leq \alpha,$$

where α is some fixed small number.

P_θ (reject H) is called the power function and $\sup_{\theta \in H} P_\theta$ (reject H) is called the size of the test. Hence the strategy (10.1) consists of maximizing power subject to a constraint on size. The strategy (10.1) is seen to be an attempt to control the risks of the two decisions "on balance." There are several difficulties for the decision theoretic model of hypothesis testing. If the true value model is indeed known to hold, then acting like $\tau \notin H$ is equivalent to acting like $\tau \in \Omega - H = K$, and failing to reject H is the same as accepting H. A decision is forced, whereas we might sometimes wish to reserve judgment. The distribution of data will not really be "known" to belong to a family, only conjectured. Thus there will be two other error types; we may for instance act like K is true when neither H nor K is true.

The likelihood ratio test of size α for testing θ_0 vs θ_1 is that we reject (accept) $\tau = \theta_0$ if $r(y) = f_{\theta_0}(y)/f_{\theta_1}(y) \leq k(> k)$, where k is chosen to satisfy $P_{\theta_0}(r(y) \leq k) = \alpha$. The jewel in the crown of the accept–reject theory, the Neyman–Pearson lemma, is that if H and Ω-H each consist of a single point, and the likelihood ratio is continuously distributed, then the likelihood ratio test satisfies the strategy (10.1).

10.3. Frequency Interpretation of Confidence Intervals

In the context of the true value model of an experiment, $(Y, \theta, \mathbb{F})$ a confidence interval of level 0.95 for true value τ is a random region I(Y) such that

$$P\{\theta \in I(Y) \mid \theta\} = 95\%, \theta \in \Omega.$$

The experiment is performed, y is observed, and it is declared that τ is in $I(y)$ with 95% confidence. A common interpretation of this statement is that it has the meaning that if we perform the same single instance confidence interval behavior over and over then according to the law of large numbers nearly 95% of our statements will be correct.

A criticism is that this interpretation does not explain the confidence interval behavior since a statistician does not go through life continually constructing confidence limits of the same level for the same parameter of repetitions of the same experiment. But this can be fixed. For a sequence of confidence intervals, I_1 of level $1 - \alpha_1$ for parameter θ_1 of experiment E_1, I_2 of level $1 - \alpha_2$ for parameter θ_2 of experiment E_2, etc., let S_n/n be the proportion of the first n which cover their true parameters. Theorem 5.4 concludes $S_n/n \rightarrow 1 - \bar{\alpha}$ since $V(S_n /n) = \sum_{i=1}^{n} \alpha_i (1 - \alpha_i)/n^2 \leq 1/n$. If level and experiment vary, we are still assured that $100(1 -\bar{\alpha})\%$ of our confidence intervals will cover their true parameters so long as repetitions are independent.

It is reassuring that a large proportion of our confidence interval statements will be correct. But the question of usual interest is: what does, say, the first **realized** interval tell us about τ_1, the first true parameter?

Barnett (1982, p. 35) comments,

(behavioral inference) provides little comfort...to the interested party in any practical problem where a single inference is to be drawn with no obvious reference to a sequence of similar situations.

He presents a practical example in which the proportion defective θ, of a batch of components is of interest and inferences are to be based on the proportion defective, $\tilde{\theta} = R/n$, in a sample of n components.

...the sampling distribution of $\tilde{\theta}$ ... must be viewed as having as 'collective' (in vonMises terms) the set of values of $\tilde{\theta}$ which might arise from repeated random samples of n components drawn from (then replaced in) the current batch.

Barnett (1982, p. 34)

Barnett's exposition concludes that while it may be satisfying to know that a large proportion of the conclusions drawn from consecutive samples would be correct, this provides little comfort as far as the present isolated conclusion is concerned. For example, consider a 95% confidence interval.

All we know is that in the long run 95 percent of such intervals obtained in similar circumstances to the present one will contain θ; 5 percent will not. We have no way of knowing into which the present falls!

Barnett (1982, p. 37)

On the other hand, a common newspaper phrase, intended to lend credibility, is "according to a usually reliable source . . . ." A 95% confidence interval is a usually reliable source where we have improved on the newspaper phrase by quantifying the reliability of the source. Frequency probability does provide comfort concerning the credibility of an isolated fact. In fact, if we were to bet on the truth of the statement, we would want to offer 19:1 odds for its truth; this is relevant to the single instance. The dependability of an isolated finding is judged by the long run frequency of correct pronouncement of the procedure which produced the finding.

The Neyman–Pearson and Wald theories adopt a behavioral attitude, contemplating the taking of actions or the making of decisions. The theory then asks what behavior would prove reasonable for a long sequence of analyses of experiments and provides a nice solution based on the law of large numbers. Hypothesis tests (the behavioral practice, to be distinguished from significance tests, an evidential strength approach to be illustrated in later chapters) control the long-run frequencies of two kinds of error and the justification for confidence intervals is that repetitions of the procedure will produce intervals covering the true parameter a prescribed proportion of the time. The Neyman–Pearson–Wald behavioral attitude is that we should conclude in a single instance whatever would prove desirable in the long run.

11

A Critique of p-Values[1]

11.1. The Context

The use of significance tests to express statistical evidence has a long history.

Example 11.1 Male and female births are not equally likely

The earliest significance test which we have found, in Todhunter (1949, p. 197), is due to Dr. John Arbuthnot. About 1710, Arbuthnot observes that for 82 years the London records show more male than female births. Assuming that the chances of male and female births are equal, Arbuthnot calculates the probability of more male than female births for 82 years in succession to be $1/2^{82}$. This probability is so small that he concludes the chances of male and female births not to be equal. Arbuthnot was making a scientific inference on the basis of what we would now call an observed significance level or p-value.

Observed significance levels still seem to be the concept of choice among practitioners for measuring statistical evidence, but they have been much criticized, as in Morrison and Henkel (1970) and Royall (1997). No doubt the p-value concept can be and has been inappropriately applied, and yet—where appropriate—it serves an important function not addressed by other statistical methods.

Appropriate p-value methods should prove particularly useful in helping to resolve disagreements among the members of the group, where—as explained by Barnard (1980) and Seidenfeld et al (1989)—personal probability methods don't apply, since different members will have different prior beliefs. Seidenfeld et al. show that two Bayesians with different probabilities and utilities have no Bayesian Pareto compromises except to adopt one or the other's approach entirely. Alternatively, see our Section 9.4.

Some authors recognize two traditions of testing: (i) Neyman–Pearson–Wald hypothesis testing, which decides among behavioral responses to a performed experiment, based on the long-run error rates of the rules used for the decision;

---

[1]This chapter follows Thompson (2006), a paper written previously.

and (ii) Fisherian significance testing, which considers p-value as an expression of experimental evidence (see Lehmann, 1993).

We briefly discussed the behavioral approach in Chapter 10, an alternative recognized source is Lehmann (1986). On page 70 Lehmann introduces p-values as an accidental and tangential by-product of the Neyman–Pearson theory: "In applications, there is usually available a nested family of rejection regions, corresponding to different significance levels. It is then good practice to determine not only whether the hypothesis is accepted or rejected at the given significance level, but also to determine the... p-value.... This number gives an idea of how strongly the data contradict the hypothesis...." But Lehmann explains neither why it is good practice to present the p-value nor why p-value gives an idea of how strongly the data contradict the hypothesis. Neyman himself maintains (without explanation) that, from his behavioral hypothesis testing point of view, Fisherian significance tests do not express evidence; see Neyman, 1982, p. 1; Fisher, 1956, p. 100; and Johnston (1986, p. 491).

Significance testing has been exposited and enlarged by Kempthorne and Folks (1971) and Cox and Hinkley (1974). The major difference between significance and hypothesis testing is the important issue of interpretation. Our treatment of evidence loosely follows the significance testing tradition. To have a name we call our explanatory model "show-me" evidence. This name—despite its suggestion of parochialism— describes, rather well, the spirit which we intend.

A reason that p-values have proved so resistant to criticism is that they augment other views of scientific method; p-values conform to one scientific view of how theoretical models should be checked. Scientific knowledge consists of statements which are to be accepted, for the time being, as applicable to the conditions of an experiment or observation—essentially the unrefuted journal articles of a discipline. Scientific theories become accepted through a social process of evolution of the working hypotheses of scientists. Hence scientific process is neither personal nor behavioral but public; the central issue is not what to decide but how does one scientist convince another of his/her conclusions? Our basic tenet throughout is that *data which agrees with prediction from a theory, but disagrees with what would be predicted from current background knowledge if the theory were not true, constitutes evidence for a theory.* Writers arriving at similar views are Kuhn (1962), Tukey (1960), Toulmin (1972), Box (1980, p. 383), and Hull (1990); see our Chapter 3.

Traditional significance tests present p-values as a measure of evidence *against* a theory. But scientists wish to accept theories for the time being, not just reject them. We are more interested in evidence *for* a theory. We find that the efficacy of a p-value for this purpose depends on specifics. Our experience is that all successful explanations of evidence—including Bayesian and likelihood versions; see equation (9.1)—are relative to an alternative; therefore tests not formulated in this way—such as the test of normality—are not recommended.

Section 11.2 considers testing a simple hypothesis relative to a simple alternative and concludes that a single p-value does not measure evidence for a simple hypothesis, but consideration of both p-values of the likelihood ratio test leads to a satisfactory theory. Section 11.3 considers the consequences of this conclusion for arbitrary, perhaps composite hypotheses; a completely general solution is not obtained because dual objectives may be contradictory. Section 11.4 specializes these considerations to the case of testing a direction relative to the opposite direction and obtains an appealing solution. For one-sided tests, a single p-value does provide an appealing measure of best evidence for a theory. Section 11.5 shows that the bivariate concept of evidence has appealing asymptotic properties for our two main classes of application. Section 11.6 considers an extension of show me evidence and illustrates this idea on an important practical problem arising in safety analysis. The final section provides concluding remarks.

11.2. Simple Hypotheses

Show-me evidence is concerned with an experiment E having random outcome Y. Let y denote a possible value which Y may assume and S, the set of all such values. Initially consider only two densities, f_η and f_κ, as possible predictors of E. Allowing the symbol κ to perform double duty, the hypothesis to be tested is that f_κ will predict the outcome of E adequately for the purpose at hand and at least as well as f_η. Hypothesis κ is similarly defined.

To formulate the concepts of evidence against η and evidence for κ we first lay down a serial order to which we then assign a numerical meaning. This is the method of fundamental measurement in physics, as in Hempel (1952, p. 62).

Assumptions for the case of simple hypotheses are—

i′. For any two densities f_η and f_κ, there is available a test statistic $T = t(Y)$ which orders that sample points of S according to their agreement with η (relative to κ). That is, $t(y) < t(y')$ may be read "the agreement of y' with η (relative to κ) is greater then that of y."

ii′. The p-values or observed significance level,

$$p(y) = pr_\eta \{t(Y) \le t(y)\}.$$

is a metrization of the above ordinal scale, a measure of agreement of y with η (relative to κ); smaller values indicating poorer agreement.

A convenient general notion for the distribution of statistic T given parameter value θ is $F_{T,\theta}(t) = pr_\theta(T \le t)$. Hence $p(y) = F_{T,\eta} \{t(y)\}$.

The use of quantiles to metrize ordinal quantities is commonplace in other similar matters and would not seem to warrant comment except that it has been questioned, as in Section 4.4.3 of Berger and Wolpert (1984). As a measure of scholastic achievement, we speak of graduating in the upper ten percent of ones class, and we interpret an income below the lower quartile as an indication of poverty.

It is important to note three closely related concepts here: (i) the significance level of a performed experiment yielding data y_o is $p(y_o)$; (ii) the function $p(y)$ is a prepared scale of agreement between possible experimental outcomes, $y \in S$, and hypothesis; and (iii) the random significance level of a contemplated experiment is a random variable $P = p(Y)$.

Determining the properties of a significance level entails determining the probabilistic properties of the random significance level. All of these properties are embodied in the distributions $F_{p,\theta}(p) = pr_\theta(P \leq p)$—where $\theta = \eta, \kappa$—of the random variable $P = p(Y)$ and hence we become interested in those distributions.

There remains the crucial issue of choice of a specific test statistic. In choosing among significance procedures, it is small observed significance which corresponds to poorer agreement with η, hence, other things being equal, we prefer those procedures which produce small observed significance under hypothesis κ. But we cannot attain this goal in any deterministic way since observed significance is a realization of a random variable.

One criterion for choice of test statistic involves stochastic ordering. A statistic U is said to be stochastically smaller than V if $pr(U \leq a) \geq pr(V \leq a)$ for all a. Dempster and Schatzoff (1965) and Kempthorne and Folks (1971, Ch. 12) present an optimality theory, in terms of stochastic ordering of significance levels.

Theorem 11.1 If the η distribution of the likelihood ratio has an inverse then when hypothesis κ holds, the likelihood ratio p-value is stochastically smaller than that based on any other test statistic.

As a preliminary we need the result that—

$$pr_\eta(P \leq p) \leq p. \tag{11.1}$$

To see this, consider a random variable T with probability law pr and distribution function $F(t) = pr(T \leq t)$. Let $P = F(T)$. $Wp = \{t : F(t) \leq p\}$ and $b = \sup Wp$. $pr(P \leq p) = pr(Wp)$ and $F(b - o \leq pr(Wp) \leq F(b)$. If $F(b) \leq p$ then $pr(Wp) \leq F(b) \leq p$. If $F(b) > p$, then $pr(Wp) = F(b - o) \leq p$. In either case $pr(P \leq p) \leq p$.

Proof of Theorem 11.1

We are concerned with two statistics, $R = r(Y)$ and $T = t(Y)$, with probability distributions $F_{R,\theta}(r)$ and $F_{T,\theta}(t)$ where $\theta = \eta, \kappa$. Let $P_R = F_{R,\eta}(R)$, $P_T = F_{T,\eta}(T)$; we need to show that $pr_K(P_R \leq p) \geq pr_K(P_T \leq p)$ for $o \leq p \leq 1$.

Let $Wp = [y : F_{R,\eta}\{r(y)\} \leq p]$ and $W'p = [y : F_{T,\eta}\{t(y)\} \leq p]$; $Wp = \{y : r(y) \leq Cp\}$ where $Cp = F_{R,\eta}^{-1}(p)$. Now $pr_\eta\{P_R \leq p\} = pr_\eta(Wp) = pr_\eta\{r(Y) \leq C p\} = F_{R,\eta}\{F_{R,\eta}^{-1}(p)\} = p$ and from (11.1), $pr_\eta(W'p) = pr_\eta(P_T \leq p) \leq p$. Therefore $\int_{Wp} f_\eta(y)dy = pr_\eta(Wp) = p \geq pr_\eta(P_T \leq p) = \int_{W'p} f_\eta(y)dy$ and $\int_{Wp-W'p} f_\eta(y)dy \geq \int_{W'p-Wp} f_\eta dy$. But in Wp, $f_\eta/f_\kappa \leq Cp$ and in the complement $f_\eta/f_\kappa \geq Cp$. Therefore $\int_{Wp-W'p} Cp f_\kappa(y)dy \geq \int_{Wp-W'p} f_\eta(y)dy \geq \int_{W'p-Wp} Cp f_\kappa(y)dy$.

Dividing through by C_p and adding $\int_{Wp \cap W'p} f_\kappa(y)dy$, we obtain $\int_{Wp} f_\kappa(y)dy \geq \int_{W'p} f_\kappa(y)dy$, which is the required result.

This theorem, essentially a reinterpretation of the Neyman–Person Lemma, motivates taking the likelihood ratio, $r(y) = f_\eta(y)/f_\kappa(y)$, as an ordinal criterion of agreement of data with η(relative to κ). This choice is so intuitive that perhaps we would question any theory yielding some other order. But while stochastic ordering yields a satisfactory theory for continuous distributions, where the inverse exists, Stone (1960) observes that difficulties are encountered for discrete distributions.

Another way of supporting the likelihood ratio *p*-value, which works for discrete as well as continuous distributions, is to consider expected significance level. Let P_T denote significance level when the criterion T is used. Thompson (1985) proves the following:

Theorem 11.2 If R is the likelihood ratio statistic and T is any other left-tailed minimum sufficient statistic, then (i) $E_\eta(P_T) = E_\eta(P_R)$ and (ii) $E_\kappa(P_T) \geq E_\kappa(P_R)$.

Proof
First notice that since $R = r(Y)$ and $T = t(Y)$ are both minimally sufficient, each is a function of the other and hence $t(y) = t(y')$ if and only if $r(y) = r(y')$. Next observe that, for $\theta = H, K$,

$$E_\theta(P_T) = \int pr_H \{t(Y) \leq t(y)\} f_\theta(y)d\mu(y) = pr(T_1 \leq T_2),$$

where T_1 has the distribution of $t(Y)$ under H, T_2 has the distribution of $t(Y)$ under θ and the two are independent. To prove (i): $E_H(P_T) = pr(T_1 \leq T_2)$, where T_1 and T_2 have the same distribution, that of $t(Y)$ under H. Thus,

$$E_H(P_T) = \frac{1}{2} + \frac{1}{2} pr(T_1 = T_2) = \frac{1}{2} + \frac{1}{2} \sum_{t=-\infty}^{\infty} \{pr_H(y : t(y) = t)\}^2,$$

which is the same for all minimal sufficient statistics $t(Y)$. The above summation notation is meaningful since all but a countable number of its terms are zero. To prove (ii), write

$$E_K(P_T) - E_K(P_R) = pr(T_1 \leq T_2) - pr(R_1 \leq R_2)$$
$$= pr(T_1 < T_2, R_2 < R_1) - pr(T_1 > T_2, R_1 < R_2)$$
$$+ pr(T_1 \leq T_2, R_1 \leq R_2) - pr(T_1 \leq T_2, R_1 \leq R_2)$$
$$+ pr(T_1 = T_2, R_2 < R_1) - pr(T_1 > T_2, R_1 = R_2)$$
$$= \int\int_{\substack{t(y_1) < t(y_2) \\ r(y_2) < r(y_1)}} [f_H(y_1)f_K(y_2) - f_H(y_2)f_K(y_1)]d\mu(y_1)d\mu(y_2) \geq 0.$$

Accepting the likelihood ratio as test statistic, assumptions I' and ii' become more specific:

i.   For any two densities f_η and f_κ, the likelihood ratio $r(y) = f_\eta(y)/f_\kappa(y)$ orders the sample points of S according to their agreement with η (relative to κ). This is not an assumption about densities; in particular it is not a monotone likelihood ratio assumption. For any two densities it defines an order relation on S.

ii.

$$p(y; \eta, \kappa) = pr_\eta \left\{ \frac{f_\eta(Y)}{f_\kappa(Y)} \leq \frac{f_\eta(y)}{f_\kappa(y)} \right\} \tag{11.2}$$

is a metrization of the above ordinal scale, a measurement of agreement of y with η (relative to κ).

The corresponding measure of agreement between y and κ relative to η, is

$$p(y; \kappa, \eta) = pr_\kappa \left\{ \frac{f_\eta(Y)}{f_\kappa(Y)} \geq \frac{f_\eta(y)}{f_\kappa(y)} \right\}$$

with large values indicating greater agreement.

In the context of significance testing, Fisher (1949, p. 16) suggests that falsification is the primary mechanism for scientific learning; so, his concern is with evidence *against* a theory. We accept his measure of evidence against a theory.

iii. Data y_0 which disagrees with η relative to κ constitutes evidence *against* the predictive capacity of η; the p-value $p(y_0; \eta, \kappa)$ measures the strength of such evidence.

But we may object, as did Berkson (1942), that the conclusions of science are to be accepted, for the time being, not just not rejected. Hence the utility of the p-value as a measure of evidence *against* a theory is questionable; we are more concerned with evidence for a theory.

Thus a small value of $p(y_0; \eta, \kappa)$ constitutes evidence *against* η; but, for two reasons, it is not evidence *for* κ. First, $p(y_0; \kappa, \eta)$ may also be small, so that evidence against κ is also strong. In addition to small $p(y_0; \eta, \kappa)$, evidence *for* κ also requires that $p(y_0; \kappa, \eta)$ is moderately large. Second, there may be a third possibility, besides η and κ, which has not been considered. Evidence for κ is relative to background knowledge, where the assumption is that η and κ are the only possibilities. Not only must the data pose a problem for hypothesis η but it must at least suggest how the problem might be solved; by adopting hypothesis κ. How we might follow the basic tenet of Section 11.1 in the context of statistical theory is suggested by the notions of "size" and "type II error" in formal Neyman–Pearson hypothesis testing. But the latter notion, and that of "power," do not carry over so simply to significance testing. We augment traditional significance testing with a measure of how well data agree with alternative hypotheses of interest.

iv. Data y_0 which disagrees with η relative to κ and agrees with κ relative to η constitutes evidence *for* κ. The strength of such evidence is measured jointly by

$p(y_0; \eta, \kappa)$ and $p(y_0; \kappa, \eta)$. Strong evidence *for* κ requires the first coordinate of the vector $ev(y_0) = \{p(y_0; \eta, \kappa), \ p(y_0; \kappa, \eta)\}$ to be small and the second large.

Note that consideration of $ev(y_o)$ as a measure of evidence is conceptually different from size and power. A striking difference is that the size and power of a test can be calculated without data.

While the above structure may appear unduly complicated, it is frequently quite simple.

Example 11.2 Binomial-simple hypotheses. For the binomial probabilities

$$f_\theta(y) = \binom{n}{y} \theta^y (1 - \theta)^{n-y}, \quad \theta = \eta, \kappa$$

where $\eta > \kappa$, the inequality $f_\eta(Y)/f_\kappa(Y) \le f_\eta(y)/f_\kappa(y)$ becomes $Y \le y$. Hence

$$ev(y_0) = \left\{ \sum_{y=0}^{y_0} f_\eta(y), \ \sum_{y=y_0}^{n} f_\kappa(y) \right\}$$

Example 11.3 Jack–queen contact. Realistic instances of testing a simple hypothesis against a simple alternative are rare, but they do occur. One example is the following. The theoretical probability that at least one jack is adjacent to one queen in a shuffled deck of ordinary playing cards is 0.486 (Thompson, 1969, p. 24). To check this computation an experiment is performed. The experimenter reports that 111 shuffles out of 240 resulted in *no* jack–queen contact. This leads to 0.538 as the estimated probability of a jack–queen contact. How strong is the evidence that the experimenter mistakenly reported the number of shuffles resulting in *at least one* jack–queen contact?

In either case the experimenter is observing the number of successes Y in 240 Bernoulli trials with success probability θ. The reporting error hypothesis is that success means a shuffle with at least one jack–queen contact, and hence $\theta = 0.486$. The hypothesis of correct reporting is that success consists of a shuffle with no jack–queen contact and hence $\theta = 0.514$. As in Example 11.3, the evidence that there was an error in reporting of this specific kind is $ev(111) = \{pr_{0.514}(Y \le 111), \ pr_{0.486}(Y \ge 111)\} \simeq (0.05, \ 0.76)$. The data is consistent with the reporting error hypothesis and is difficult to explain from background knowledge by other means.

11.3. Composite Hypotheses

In generalization of the theory for simple hypotheses we now consider a class $\{f_\theta : \theta \in \Omega\}$ of densities as potential predictors of E, and a partition of Ω into two disjoint subsets, H and K. Again allowing the symbol H to perform double duty, the hypothesis H is that some member of a class of probability densities $\{f_\theta : \theta \in H\}$ will predict the outcome of E adequately for the purpose at hand and at least as well as any other density f_θ with $\theta \in \Omega$. Hypothesis K is similarly defined. The composite test assumption is that current background knowledge and the theory to be tested can be formulated as hypotheses Ω and K, respectively; $H = \Omega - K$, called the null hypothesis, is then background knowledge with K false.

Assumptions (i) and (ii) are as before except that they now apply to every pair, $\eta \in H$ and $\kappa \in K$. With regard to generalizing (iii) and (iv), if $p(y; \kappa, \eta) \geq \beta$, then we say that y agrees with κ at level β relative to η and if $p(y; \eta, \kappa) \leq \alpha$, then y disagrees with η at level α relative to κ. Now relative to κ, $\sup_{\eta \in H} p(y_0; \eta, \kappa)$ is the level of disagreement of y with H, since $\sup_{\eta \in H} p(y_0; \eta, \kappa) \leq \alpha$ if and only if y disagrees at level α with all $\eta \in H$. Similarly, $\inf_{\eta \in H} p(y_0; \eta, \kappa)$ is the level of agreement of y with κ relative to H.

Assumption (iii) becomes—

iii. Data y_0 which, relative to some $\kappa \in K$, disagrees with every f_η, $\eta \in H$ constitutes evidence *against* the predictive capacity of H; the p-value

$$p(y_0, H, K) = \inf_{\kappa \in K} \sup_{\eta \in H} p(y_0; \eta, \kappa) \tag{11.3}$$

measures the strongest of such evidence.

While we accept $p(y_0, H, K)$ as a valid measure of evidence against H, we continue to question the ultimate practical utility of any such measure; scientists are concerned with accepting theories, for the time being, not rejecting them.

The generalization of assumption (iv) is less explicit; we know how to recognize predictive evidence for a theory but there is no universally valid closed form for its measure:

iv. Data y_0 which agrees with some $\kappa \in K$ (relative to all $\eta \in H$) but disagrees (relative to κ) with all $\eta \in H$, constitutes evidence **for** K. For each $\kappa \in K$

$$ev(y_0, \kappa) = \left\{ \sup_{\eta \in H} p(y_0; \eta, \kappa) , \inf_{\eta \in H} p(y_0; \kappa, \eta) \right\} \tag{11.4}$$

measures the strength of that evidence for K.

Thus we examine (11.4) for various $\kappa \in K$. If for some $\widehat{\kappa}$, the first coordinate is small while the second is large then we have evidence *for* K measured by $ev(y_0, \widehat{\kappa})$. The data simultaneously poses a problem for hypothesis H and suggests how the problem can be resolved—by accepting hypothesis $\widehat{\kappa} \in K$. We would like to present the best such evidence for K but we are prevented from doing so by the bivariate nature of the objective function. Nevertheless, in some cases, we can carry out this optimization; this exposition focuses on those cases. Then we write $ev(y_o) = ev(y_o, \widehat{\kappa})$. for the best evidence for K.

Example 11.4 Coplanar planetary motion. A famous scientific inference based on a p-value is due to Daniel Bernoulli (see Todhunter, 1949, p. 223). Bernoulli's 1734 computation amounts to the following. For our solar system, as know in Bernoulli's time, the greatest inclination of the plan of any of six orbits to the sun's equator is $7° 30'$ or $1/12$ or $90°$. Bernoulli takes $1/12^6$ to be the probability that each of six inclinations would be so small. This small probability, along with several similar computations, convinced Bernoulli that we cannot attribute to chance the small mutual inclinations of the planetary orbits.

A modern way of viewing Bernoulli's problem and solution is to introduce the additional structure that the six measured angles $Y_1, \ldots, Y_6$ are a sample from a uniform distribution on the interval zero to $\theta (0 < \theta \leq \pi/2)$ and that we are measuring evidence against the hypothesis H, that $\theta = \pi/2$ and for the hypothesis K that $0 < \theta < \pi/2$. Writing $T = \max (Y_1, \ldots, Y_6)$ and $t = \max(y_1, \ldots, y_6)$ the density is

$$f_\theta(y) = \begin{cases} \theta^{-6}, & 0 \leq t \leq \theta \\ 0, & \theta < t \end{cases}$$

which results in the likelihood ratio

$$\frac{f_{\pi/2}(y)}{f_\kappa(y)} = \begin{cases} \left(\dfrac{2\kappa}{\pi}\right)^6, & 0 \leq t \leq \kappa \\ \infty, & \kappa < t \end{cases}$$

for $\kappa < \pi/2$. Thus

$$p(y; \pi/2, \kappa) = \begin{cases} 1, & \kappa < t \\ \left(\dfrac{2\kappa}{\pi}\right)^6, & 0 \leq t \leq \kappa \end{cases}$$

and

$$p(y; \kappa, \pi/2) = \begin{cases} 0, & \kappa < t \\ 1, & 0 \leq t \leq \kappa. \end{cases}$$

For given y, the smallest value of $p(y; \pi/2, \kappa)$ and the largest value of $p(y; \kappa, \pi/2)$ both occur at $\hat{\kappa} = t$. Hence from (11.4), for Bernoulli's data, $t_0 = 7° 30'$, the best evidence for $\theta < \pi/2$ relative to $\theta = \pi/2$ is $(12^{-6}, 1)$, augmenting Bernoulli's computation.

Example 11.1 (continued) Arbuthnot's test illustrates a two-sided problem. Writing θ for the probability of a male birth, the two hypotheses are $H : \theta = 1/2$ and $K : \theta \neq 1/2$. The expression (11.4) is

$$ev(82, \kappa) = \begin{cases} (2^{-82}, 1), & \kappa > 1/2 \\ (1, \kappa^{82}), & \kappa < 1/2 \end{cases}$$

taking $\hat{\kappa} > 1/2$, the evidence that male and female births are not equally likely is $ev(82, \hat{\kappa}) = (2^{-82}, 1)$, which is quite strong.

11.4. Test of a Direction

Example 11.5 Vitamin C. Consider, as did Berger and Berry (1988), a "double-blind" experiment conducted to study the effectiveness of vitamin C in treating the common cold and suppose that 17 matched pairings of vitamin C treatment with a placebo (P) have been obtained. Let θ denote a fitted probability that, in a single pairing, C does better than P and let Y denote the random number of pairs in which C beats P. Background assumptions are

$$f_\theta(y) = \binom{17}{y} \theta^y (1-\theta)^{17-y}, y = 0, \ldots, 17, \quad 0 \le \theta \le 1.$$

The novel theory K is that vitamin C is effective—that $\theta > \frac{1}{2}$; H is that $\theta \le \frac{1}{2}$. Hence, for $\eta \in H$ and $\kappa \in K$, $\eta \le 1/2 < \kappa$, $p(y; \kappa, \eta) = pr_\eta(Y \le y)$ and $p(y; \kappa, \eta) = pr_\kappa(Y \ge y)$. Now $ev(y, \kappa) = \{pr_{1/2}(Y \le y), pr_\kappa(Y \ge y)\}$ and, optimizing over $K.ev(y) = \{pr_{1/2}(Y \le y), pr_{1/2}(Y \ge y)\}$.

If $Y = 4$ is observed, then $ev(y) = (0.025, 0.994)$. The predictive evidence that vitamin C is effective—that $\theta > 1/2$—is that the success of the K prediction $Y \ge 4$ is difficult to explain if $\theta \le 1/2$ but is easily explained if $\theta > 1/2$.

Note that in this example, the best bivariate evidence will be strong whenever the p-value is small, hence agreement of data with K need not be checked separately. While this is not true in general, we will show in Theorem 11.4 that it does hold for a class of one-sided tests.

It is a usual feature of significance tests that they serve multiple purposes. The p-value $p(y; \theta_0, \theta_1)$ is to measure strength of predictive evidence against θ_0 relative to the alternative θ_1. But often $r(y) = f_{\theta_0}(y)/f_{\theta_1}(y)$ and $f_{\theta_0}(y)/f_\kappa(y)$ yield the same order on S and hence the same p-value. In this case we say that θ_1 and κ are in the same direction from θ_0. Let K be the set of all such $\kappa \in \Omega$. Further, if $f_\eta(y)/f_{\theta_0}(y)$ yields the same order as $r(y)$, then we say that η and θ_1 are in opposite directions from θ_0; let H denote θ_0 augmented with all such η. Now writing $R = r(Y)$, since $f_\eta/f_\kappa = (f_\eta/f_{\theta_0})(f_{\theta_0}/f_\kappa)$, $p(y; \eta, \kappa) = F_{R,\eta}\{r(y)\}$ and $p(y; \kappa, \eta) = pr_\kappa \{R \ge r(y)\}$ for all $\eta \in H$ and $\kappa \in K$. Note that $p(y; \kappa, \eta)$ does not depend on η. Also

$$\max_{\eta \in H} p(y; \eta, \kappa) = pr_{\theta_0}\{R \le r(y)\} = p(y; \theta_0, \theta_1) \tag{11.5}$$

In fact, $F_{R,\eta}(a) \le F_{R,\theta_0}(a)$ for all $\eta \varepsilon H$ and all a. For if not, then $F_{R,\eta}(a) > F_{R,\theta_0}(a)$ and also $1 - F_{R,\eta}(a_0) < 1 - F_{R,\theta_0}(a)$ Then for some x and y, $r(x) \le a_0 < r(y)$, but $f_\eta(x)/f_{\theta_0}(x) > 1 > f_\eta(y)/f_{\theta_0}(y)$ in contradiction with $\eta \varepsilon H$. The argument here is reminiscent of monotone likelihood ratio theory, but we make no such assumption.

Equations (11.5) and (11.3) establish—

Theorem 11.3 Writing K for the direction of θ_1 from θ_0 and H the opposite direction, we have $p(y; \theta_0, \theta_1) = p(y; \theta_0, \kappa)$ for every $\kappa \in K$ and in fact $p(y; \theta_0, \theta_1) = p(y; H, K)$.

Theorem 11.3 makes it clear that a small p-value by itself does not constitute evidence for θ_1; for κ in the direction of θ_1, $p(y; \theta_0, \theta_1) = p(y; \theta_0, \kappa)$ constitute equal evidence *against* θ_0 but y may agree better with κ then with θ_1. The mathematical starting point for significance tests is that of simple hypothesis versus simple alternative. But whether intended or not the same p-value measures strength of predictive evidence against a direction. The basic significance test is the test of a direction against the opposite direction, it becomes simple only if Ω is restricted by assumption to only two points.

We now turn to evidence *for* K, the direction of θ_1. From (11.4) and (11.5), $ev(y, \kappa) = \{p(y; \theta_0, \theta_1), p(y; \kappa, \theta_0)\}$. Since the first coordinate does not depend on η, we may optimize to obtain

$$\{p(y; \theta_0, \theta_1), \sup_{\kappa \varepsilon K} p(y; \kappa, \theta_0)\} \tag{11.6}$$

as our bivariate measure of evidence for the direction of θ_1.

Now consider $Y = (Y_1 \ldots, Y_n)$ to be a sample from the single-parameter exponential family in natural parameter form; Y will have density of the form

$$f_0(y) = \exp \{\theta \Sigma s(y_i) + nc(\theta) + \Sigma b(y_i)\} \tag{11.7}$$

Examples are the normal, gamma, binomial, and Poisson distributions. A sufficient statistic for θ is $T = \Sigma s(Y_i)$. We have that the agreement of y with θ_0 relative to θ_1 is greater then that of zif and only if $(\theta_0 - \theta_1)\{\Sigma s(y_i) = \Sigma s(z_i)\} > 0$. Thus in this case, "direction" has its usual geometric meaning in the natural parameter space; θ is the same "direction" from θ_0 as θ_1 if and only if $(\theta_0 - \theta_1)(\theta_0 - \theta) > 0$. If $\theta_1 < \theta_0$, then $K = (\theta : \theta < \theta_0)$, $H = (\theta : \theta \geq \theta_0)$, and the evidence *against* H is $pr_{\theta_0}(T \leq t_0)$, where t_0 is the observed value of T; taking the limit under the integral sign, the evidence (11.6) *for* K is

$$\left[pr_{\theta_0}(T \leq t_0), pr_{\theta_0}(T \geq t_0) \right] \tag{11.8}$$

in generalization of the result for Example 11.5.

Theorem 11.4 When sampling from the single parameter exponential family, the best evidence for κ is given by (11.8), and hence small *p*-value alone constitutes evidence for a direction.

The test of Theorem 11.4 can be related to expectations. Write $\mu = \mu_\theta = E_\theta s(Y_1)$, $\sigma^2 = \sigma_\theta^2 = \text{var}_\theta s(Y_1)$, $\mu_0 = \mu_{\theta_0}$, $\sigma_0^2 = \sigma_{\theta_0}^2$ and   consider the moment generating function, M (z), of Y_1. We have $M(z) = \exp\{c(\theta) - c(z + \theta)\}$, $\mu = M'(0) = -c'(\theta)$ and $\sigma^2 = -c''(\theta)$. Since $c''(\theta) < 0$, μ is an increasing function of θ, $K = \{\mu : \mu < \mu_0\}$ and (11.8) measures evidence for $\mu < \mu_0$.

11.5. Asymptotic Considerations

First, consider the large sample behavior of the *p*-value of Theorem 11.4. From the central limit theorem, $p(y; H, K) = pr_{\theta_0}(T \leq t_0) \sim \Phi \left(\dfrac{t_0 - n\mu_0}{n^{1/2} \sigma_0} \right)$ and

$$pr \{p(Y, H, K) \leq p\} \sim pr_\theta \left\{ \Phi \left(\frac{T - n\mu_0}{n^{1/2} \sigma_0} \right) \leq p \right\}$$

$$\sim \Phi \left\{ \frac{\sigma_0 \Phi^{-1}(p) + n^{1/2}(\mu_0 - \mu)}{\sigma} \right\}.$$

Theorem 11.5 When testing the direction $K = \{\mu : \mu < \mu_0\}$ versus the opposite direction, for the single parameter exponential family, a good approximation to the median of $p(Y; H, K)$ is $m(\mu) = \Phi\left\{n^{1/2}(\mu - \mu_0)/\sigma_0\right\}$. Note that $m(\mu)$ is independent of σ and has the "right" asymptotic limits as $n \to \infty$.

Also, with probability one as $n \to \infty$, the evidence (11.8) for $\mu < \mu_0$ approaches the ideal values of (0,1) when it is true, and (1,0) when $\mu > \mu_0$; for $\mu = \mu_0$ we have $\theta = \theta_0$, $\sigma = \sigma_0$, and p(Y;H,K) is uniformly distributed on the unit interval.

For a second asymptotic investigation, return to the simple hypothesis situation of Section 11.2, where we are measuring evidence *for* κ relative to η. Restrict consideration to sampling from the exponential family (11.7) where $\eta > \kappa$ and hence $\mu_\eta > \mu_\kappa$. We have

$$p(y_0, \eta, \kappa) = pr_\eta(T \le t_0) \sim \Phi\left(\frac{t_0 - n\mu_\eta}{n^{1/2}\sigma_\eta}\right) \text{ and } p(y_0, \eta, \kappa) \sim 1 - \Phi\left(\frac{t_0 - n\mu_\kappa}{n^{1/2}\sigma_\kappa}\right)$$

So

$$pr_\theta\left\{p(Y; \eta, \kappa) \le \alpha, \ p(Y; \kappa, \eta) \le \beta\right\}$$

$$\sim \begin{cases} \max\left[0, \alpha - \Phi\left\{\dfrac{\Phi^{-1}(1 - \beta)\sigma_\kappa + n^{1/2}(\mu_\kappa - \mu_\eta)}{\sigma_\eta}\right\}\right], \theta = \eta \\[4mm] \max\left[0, \Phi\left\{\dfrac{\Phi^{-1}(\alpha)\sigma_\eta + n^{1/2}(\mu_\eta - \mu_\kappa)}{\sigma_\kappa}\right\}\right], \theta = \kappa \end{cases}$$

$$\xrightarrow[n\to\infty]{} \begin{cases} \alpha, \theta = \eta \\ \beta, \theta = \kappa \end{cases}$$

Theorem 11.6 Asymptotically, under hypothesis η, $ev(Y) = \{p(Y; \eta, \kappa), p(Y; \kappa, \eta)\}$ is uniformly distributed on the line segment $\{(0, 0), (1, 0)\}$ and under κ it is uniform on $\{(0, 0), (0, 1)\}$.

While agreement with a false hypothesis is zero—the ideal value—agreement with a true hypotheses is distributed around one-half. Nevertheless, for large samples, $ev(Y)$ allows us to identify the correct hypothesis with probability one.

11.6. Intuitive Test Statistics

With slightly less theoretical justification, the range of applicability of predictive evidence can be enlarged somewhat by considering test statistics other than the likelihood ratio. Consider the special circumstance that for all $\eta \in H$, $\kappa \in K$, a common statistic $T = t(Y)$ orders the sample points of S according to their relative consistency with η over κ. Assumptions entirely analogous to those given previously lead to equations (11.3) and (11.4), where now $p(y; \eta, \kappa) = pr_\eta\{T \le t(y)\}$ and $p(y; \kappa, \eta) = pr_\kappa\{T \ge t(y)\}$. In this special case, the first

TABLE 11.1. Rocket firing data

Firing	Burning time (coded)	Tube time (same code)
1	58.671	69.524
2	61.284	69.542
3	60.619	71.256
4	60.699	69.462
5	60.101	70.404
6	58.619	70.602
7	59.426	72.732
8	60.096	70.420
9	61.389	69.528
10	61.249	71.412

coordinate of (11.4) does not depend on κ so that we may optimize over κ to obtain

$$\left[\sup_{\eta \in H} pr_\eta \{T \le t(y)\}, \sup_{\kappa \in K} pr_\kappa \{T \le t(y)\} \right] \tag{11.9}$$

as a bivariate measure of the evidence of y for K.

Example 11.6 Safety analysis. If the propellant of a shoulder fired rocket is still burning when it leaves the tube, then the operator will be burned. Given the real firing data of Table 11.1, is the rocket safe to use?

One formulation is as follows. Let D denote tube minus burning time. The probability of an accident is $\alpha = pr(D < 0)$. We might structure questions concerning the safety of the rocket in terms of α. We might, for example, require $\alpha \le 0.005$. If burning and tube time are bivariate normal, then D will be normal with unknown mean μ and variance σ^2. Now $\alpha = pr[(D - \mu)/\sigma \le -\mu/\sigma] = \Phi(-\mu/\sigma)$ where Φ is the standard normal distribution function. The requirement $\alpha \le 0.005$ is equivalent to $\mu/\sigma \ge 2.58$.

From Table 11.1, the unbiased estimates of μ and σ^2 are $\bar{d} = 10.273$ and $s_D^2 = (1.616)^2$, yielding 6.36 as an estimate of μ/σ. This corresponds to 0.00000 as an estimate for α, the probability of an accident.

We might examine the evidence that the rocket is safe, $\mu/\sigma \ge 2.58$. The likelihood ratio test indicates the sample mean as test statistic; this leads to a useless measure of evidence which depends on the unknown nuisance parameter σ^2. Instead we may consider the test statistic $T = \sqrt{10}\,\bar{D}/s_D$, which has the noncentral t-distribution with 9 degrees of freedom and noncentrality parameter $\sup_{\delta \le 8.16} P_\delta(T \ge 20.1) = 0.4\%$. The hypothesis, unsafe according to the above criterion, corresponds to $\delta \le 8.16$; large values of T are unfavorable to the hypothesis. The observed value of T is 20.1. The observed significance level is

$$\sup_{\delta \le 8.16} P_\delta(T \ge 20.1) = 0.4\%,$$

and the evidence (11.8), that the rocket is safe, is (0.4%, 99.6%), which is relatively strong. Many safety analysis applications can be formulated in terms of the difference of two random variables.

11.7. Discussion

The context is that p-values have a long tradition of usage to express statistical evidence but they have been much criticized. Criticism mostly takes three forms: (i) They fail to distinguish between evidence *for* and evidence *against* a theory (ii) They neglect alternative hypotheses. (iii) They disagree with other models of statistical inference.

We agree with the first criticism: p-values are readily explained as evidence *against* a hypothesis, but in practice we are interested in evidence for an hypothesis.

Second, as Friedman et al. (1978, p. 492) explain, failure to consider the alternative leaves the null hypothesis to "take the heat" and we may even uncritically end up with the least supported of two hypotheses. Presumably, our treatment of alternative probabilities will answer some of these complaints. However, the suprising Theorem 11.4 implies that for one important class of tests, small p-value alone constitutes evidence for a hypothesis; alternative hypothesis probabilities need not be considered separately.

The criticism that p-values disagree with other forms of inference is actually a special case of a much more general misconception—discussed in Section 2.3— about the nature of applied mathematics. We postpone detailed discussion of this criticism to Section 12.3.

We conclude as follows; the p-value can be explained as one measure of best evidence against a hypothesis, but we do not recommend that usage since experimenters are really interested in evidence for a hypothesis. In general, evidence for a hypothesis requires the additional consideration of alternative p-values, as in (iv) of Section 11.3. This is true in particular, as in (iv) of Section 11.2, for testing simple hypotheses. But, as explained in Section 11.4, for distributions in the exponential family, it turns out that small p-value alone constitutes evidence for a direction.

While show-me evidence provides an explanation for several interesting classes of problems, its applicability is limited. It may not be possible to formulate a desired hypothesis in terms of a direction, and show-me evidence has nothing to say about the handling of nuisance parameters. Section 11.7 suggests one way in which the applicability of show-me evidence might be enlarged.

12

The Nature of Statistical Evidence

12.1. Introduction

As discussed in Chapter 8, Birnbaum introduces $E_v(E, y)$, the evidential meaning of obtaining data y as an instance of experiment E. Following Birnbaum, various authors have wrestled with the problem of developing a single set of postulates under which statistical inference can be made coherent. But as we claim in Section 8.3, $E_v(E, y)$ does not exist. Evidence is grounds for belief—an imprecise concept. There must be many valid reasons for believing and hence many ways of making the evidence concept precise. Most of our beliefs are held because mother—or someone else we trust—told us so. The law trusts sworn testimony. Scientific and statistical evidence are other different grounds for belief—supposedly particularly reliable kinds. Instead of $E_v(E, y)$ we are concerned with $E_v(E, T, y)$, the evidential meaning of observing y as an instance of E, in the context of theory T.

What can we expect of a theory of statistical inference? We can expect an internally consistent explanation of why certain conclusions follow from certain data. The theory will not be about inductive rationality but about a *model* of inductive rationality. Statisticians are used to thinking that they apply their logic to models of the physical world; less common is the realization that their logic itself is only a model. Explanation will be in terms of introduced concepts which do not exist in nature. Properties of the concepts will be derived from assumptions which merely seem reasonable. This is the only sense in which the axioms of any mathematical theory are true; see Chapter 2 and Wilder (1983, Section 1.5). We can expect these concepts, assumptions, and properties to be intuitive but, unlike natural science, they cannot be checked by experiment. Different people have different ideas about what "seems reasonable," so we can expect different explanations and different properties. We should not be surprised if the theorems of two different theories of statistical evidence differ. If two models had no different properties then they would be different versions of the same model (see Chapter 2 for greater detail). We should not expect to achieve, by mathematics alone, a single coherent theory of inference, for mathematical truth is conditional and the assumptions are not "self-evident." Faith in a set of assumptions would be needed to achieve a single coherent theory.

12.2. Birnbaum's Theorem

In the context of the true value model of Section 8.1, Birnbaum (1962) considers various principles for how experimental evidence should behave and obtains the likelihood principle as a consequence of the conditionality and sufficiency principles. Berger and Wolpert (1984, p. 28), and others attribute far-reaching significance to Birnbaum's theorem.

A statistic s(y) is **sufficient** for the model $(Y, \theta, \mathbb{F})$ if the conditional distribution of the data given the sufficient statistic does not involve θ. The **sufficiency principal** is that if s(y) = s(z) then $E_v(E, y) = E_v(E, z)$. The justification is that, if $\mathbb{F}$ is known to contain the true density, then once s becomes available we may regard the rest of the data as if generated by a process not involving, and therefore uninformative about, the parameter θ. All major theories of statistics adhere to the sufficiency principle. Note however, that in Chapter 8 we observe that typically the functional form of the true density will be unknown.

The likelihood principle is about the **likelihood function** $f_\theta(y)$—the density considered as a function of θ with y fixed at the observed value. The (strong) **likelihood principle** is that identical inferences, about a common unknown quantity τ, should be drawn from two different experiments resulting in proportional likelihood functions, or that all evidence about the true value τ is contained in the likelihood function. The likelihood principle is at odds with forward-looking measures of evidence such as coverage probabilities and significance levels. A standard example is that obtaining y successes in n trials, according to the binomial or negative binomial experiment, results in proportional likelihoods but different significance levels.

The likelihood principle is also at odds with established scientific practice. Wilson (1952) points out that journal articles in the experimental sciences have the standard section structure: (i) Background, (ii) Methods and Materials, (iii) Results, and (iv) Interpretation. The purpose of giving the background is to indicate the context of the article: the author's working hypotheses and any school to which he may belong, the question being asked of nature and the criteria or tradition according to which the experiment is to be judged as an answer. Section (ii) explains the experimental recipe, what is to be done and what is to be observed. Section (iii) summarizes the current performance of the experiment. The final section discusses how the puzzle came out, and particularly the class of related problems for which the results may be expected to come out the same way. Scientific practice holds that many things besides the likelihood function need to be included in the report of an experimental investigation. For example, if y successes are obtained in *n* trials, then Section (ii) should tell us whether the experiment was binomial or negative binomial.

The reason that experimental scientists want more than just the likelihood function in the report of an experiment is that, though they might talk otherwise, ASTM committees (discussed in Section 8.4) are not engaged in discovering "true values." They are concerned with constructing experimental methods which are reliable in that one can predict the results of future experiments on the basis of past experiments. Deming (1986, pp. 350–1) says it well:

Unfortunately, future experiments (future trials, tomorrow's production) will be affected by environmental conditions (temperature, materials, people) different from those that affect this experiment. It is only by knowledge of the subject matter, possibly aided by further experiments to cover a wider range of conditions that one may decide, with a risk of being wrong, whether the environmental conditions of the future will be near enough the same as those of today to permit the use of results in hand.

The likelihood principle is a very nice property; too good to be true in general. We don't need to know what experiment we are analyzing or what assumptions went into the analysis, just the likelihood function. "Indeed if the investigator died after reporting the data but before reporting the design of the experiment" (Berger and Berry, 1988, p. 162), we could still use the results in hand to predict future experiments. Unfortunately, as Deming explains so clearly, we need a great deal more than the likelihood function. If an ASTM committee is to construct a reliable experimental method, then careful discussion will be needed—and is in fact the rule—about what **is** the relevant experiment.

Berger and Wolpert (1984, p. 28), state the conditionality principle (WCP) as follows:

Suppose there are two experiments $E_1 = (X_1, \theta, \{f_\theta^1\})$ and $E_2 = (X_2, \theta, \{f_\theta^2\})$, where only the unknown parameter θ need be common to the two experiments. Consider the mixed experiment E^*, whereby $J = 1$ or 2 is observed, each having probability $\frac{1}{2}$ (independent of θ, X_1, or X_2), and experiment E_J is then performed. Formally, $E^* = (X^*, \theta, \{f_\theta^*\})$, where $X^* = (J, X_j)$ and $f_\theta^*((j, x_j)) = \frac{1}{2} f_\theta^j(x_j)$. Then,

$$Ev(E^*, (j, x_j)) = Ev(E_j, x_j),$$

i.e., the evidence about θ from E^* is just the evidence from the experiment actually performed.

On their pages 6 and 7 they present the following example as supporting intuition for the principle.

Example 12.1 Berger and Wolpert.
(It) is the key to all that follows. Suppose a substance to be analyzed can be sent either to a laboratory in New York or a laboratory in California. The two labs seem equally good, so a fair coin is flipped to choose between them, with "heads" denoting that the lab in New York will be chosen. The coin is flipped and comes up tails, so the California lab is used. After awhile, the experimental results come back and a conclusion must be reached. Should this conclusion take into account the fact that the coin could have been heads, and hence that the experiment in New York might have been performed instead? The obvious intuitive answer... is that only the experiment actually performed should matter.

Practicing data analysts have largely ignored these three principles; But statisticians concerned with foundations cannot afford to ignore Birnbaum's (1962) theorem—that the sufficiency and weak conditionality principles imply the likelihood principle. As Berger and Wolpert explain, the likelihood principle is extremely radical from the point of view of classical statistics. Yet to reject the likelihood principle one must logically reject either sufficiency or weak conditionality. But sufficiency is itself a cornerstone of classical statistics and there is nothing in statistics as "obvious" as weak conditionality or the laboratory analysis Example 12.1. In our view there are problems with this use of Birnbaum's theorem. We do not present a proof of Birnbaum's theorem since our complaint is not with the logical rigor but with the true value formulation.

Berger and Wolpert say that the WCP "is nothing but a formalization of" Example 12.1. We argue that it is a *mis* formulation; neither the hypothesis nor the conclusion of with WCP is appropriate for Example 12.1. This doesn't negate the WCP—we take no position on the WCP—but it does cast doubt on this basis for "all the follows" and it illustrates that the absence of a true value can make a difference in our conclusions.

Applying the WCP to Example 12.1 assumes two experiments E_1 and E_2 represented as triples and observing the world directly with equal skill. Hence, from the perspective of the true parameter model, the corresponding measurements, X_1 and X_2 will have a common true density f_τ, and X_J will have density $f_\tau/2 + f_\tau/2 = f_\tau$. The three experiments consist of observing the same density so, the same observed value yields the same evidential meaning.

But we aren't free to make these assumptions: The major issue of experimentation is *whether* results are reproducible by different laboratories. f_τ and τ do not exist, experiments aren't triples and they aren't about parameters. Parameters are about experiments. The assumption that "the two laboratories seem equally good" throws the proverbial baby out with the bath water. The only way we could arrive at that conclusion—with a chance of being wrong—is on the basis of a previously conducted interlaboratory comparison of measurements.

If indeed one observation is all of the data, then Example 12.1 is a special case of interlaboratory experimentation, the subject discussed in Section 8.5. We can expect that two laboratories carrying out the same method (C,X) on the same substance are different experimental processes, $E_j = (C, X; j); j = 1, 2$. They are observing the world through different instruments, operators, and supporting theory.

The hypothesis and intuitive conclusion of Example 12.1 ignores the target population. The role of the coin flip is left uninterpreted. If, as is common in studies of interlaboratory experimentation, the target population is—in the language of 8.5—the method of experimentation rather than one of the experimental processes, and the coin flip is to reflect a hypothesis that the laboratories will be used equally often, then the mixed experiment has an additional source of variability which *should* matter in the summarizing report. Stated symbolically, the best-fitting parameter for $E_j = (C, X; j)$ is μ_j; $j = 1, 2$. The best fitting parameter for $E^* = (C, X)$ is $(\mu_1 + \mu_2)/2$. The report to be written for E_1 is that x_1 is an (unreliable) estimate of μ_1. The report to be written for E^* is that x_1 estimates $(\mu_1 + \mu_2)/2$. Note that the best fitting parameters and reports differ among experiments. $E_v(E^*, T_1, (1, x_1)) \neq E_v(E_1, T_2, x_1)$.

12.3. Counterexamples

The context of this section is that there is disagreement among experts about the nature of statistical evidence and consequently much use of one formulation to criticize another. Neyman (1950) maintains that, from his behavioral hypothesis testing point of view, Fisherian significance tests do not express evidence. Royall (1997) employs the "law" of likelihood to criticize hypothesis as well as significance testing. Pratt (1965), Berger and Selke (1987), Berger and Berry (1988), and Casella and Berger (1987) employ Bayesian theory to criticize sampling theory.

O'Hagen (1994, p. 18), and others, refer to this practice as "detecting counterexamples"; but that is not what a counterexample is. A counterexample to a theorem is an instance where the hypothesis is satisfied but the conclusion is not. The hypotheses of the various theories of inference differ. For example, Bayesian inference assumes that all relevant quantities are random, whereas sampling theory assumes that some quantities are random while others are fixed but unknown constants. Neither set of assumptions is more 'true" in the abstract; both are choices of volition and it is not clear how to choose.

Critics assume that their findings are about evidence, but they are at most about models of evidence. Many theoretical statistical criticisms, when stated in terms of evidence, have the following outline: According to model A, evidence satisfies proposition P. But according to model B, which is correct since it is derived from "self-evident truths," P is not true. Now evidence can't be two different ways so, since B is right, A must be wrong. Note that the argument is symmetric: since A appears "self-evident" (to adherents of A) B must be wrong. But both conclusions are invalid since evidence can be *modeled in different ways,* perhaps useful in different contexts and for different purposes. From the observation that P is a theorem of A but not of B, all we can properly conclude is that A and B are different models of evidence. In the notation of Section 8.3, it is clear that $E_v(E, A, y)$ need not equal $E_v(E, B, y)$. The common practice of using one theory of inference to critique another is a misleading activity.

As a particular example, consider the criticism of p-values, postponed from Chapter 11. Royall (1986), Peto et al (1976), and Lindley and Scott (1984) point out that different Bayesian meanings of significance tests depend on sample size in different ways.

On the other hand, presumably, the meaning of a given probability does not vary over experiments. That is, if $pr(A, F) = pr(B, E)$, then event A has the same tendency to happen when experiment F is performed as does event B when E is performed. If when flipping a coin pr(head) $= 1/2$ meant something different than pr(even) $= 1/2$ when rolling a die, then probability would be a useless concept. For each $y \in S$ the p-value $p(y; \eta, \kappa)$ of Chapter 11, consists of a probability theoretically calculated before data. Hence the show-me meaning of $p(y; \eta, \kappa)$, $y \in S$, and the special case $p(y_0; \eta, \kappa)$ does not vary with sample size.

This discrepancy is not surprising since significance tests do not belong to the Bayesian logic; the theorems of different models of inductive rationality can be expected to differ. This is once again the Euclidean misconception.

Berger and Wolpert (1984) and others, employ the conditionality principle to criticize classical inference and in particular p-values. But the evidence of the conditionality principle is about the true density of experimental data known to be a member of a parametric class, while show-me evidence of Chapter 11 is about whether a subset of an assumed class of densities, includes a member that fits data adequately. An experiment is a recipe for performance, it isn't about parameters; parameters are about experiments. It cannot be literally true that we know the functional form of the true density and this makes a difference.

13

The Science of Statistics

13.1. Meanings of "Science"

On the inside of the front cover of each issue of its *Journal,* the American Statistical Association declares that it is a scientific organization. In what sense is statistics scientific? There are several common usages of the word "science." First, science is sometimes used as a synonym for systematized knowledge. Or, in more detail, a science is the systematized knowledge produced by the study of the structure of a class of concepts. A second usage is that science is explanation. Third and more narrowly, science is sometimes taken to mean the systemized knowledge of "nature," of the "real world." A fourth usage, due to Karl Pearson, is that any field of study which employs the scientific method of hypothesis, deduction and experiment is a science. A fifth usage is common in physics; Ruhla (1993) writes "prediction = science."

The first concept of science is very broad; virtually all academic subjects are systemized knowledge. History is systemized knowledge and historians frequently hypothesize and deduce, but that part of history which is "just one damned fact after another" probably isn't scientific. In spite of the view that "history repeats itself," history is an unreliable predictor, for it is not subject to experimental test; the situation is not repeatable. But much history *can* be considered an explanation. Historians of the American civil war speculate: "If Stonewall Jackson had not been killed at Chancellorsville then Gettysburg would have been a small skirmish on Lee's march to Washington." Do we want history to be a science?

The latter three usages of the word science are all attempts to characterize natural science. Most discussions of science are implicitly limited to natural science, the study of nature or the world out there. Clearly, chemistry, physics, and biology are natural sciences but there are also computer science, political science, social science, and creation science.

In contrast, many worthwhile human activities are not at all scientific. While some artists have used the results of science, artistic painting would seem a prime example of a nonscientific activity. Artistic painting is not expected to follow the rules of logic and there is no corrective mechanism corresponding to the experiment in science.

With regard to the fourth concept of science, going through the motions of scientific method is not enough to imply science. Margaret Jarman Hagood (1941, pp. 425–432) states the issue for sociology with refreshing objectivity. There is—

the premise that there is a stability, a regularity, an orderliness in the occurrence of sociological phenomena, even though it is dynamic and ever-changing, and that one task of developing a scientific sociology embraces the description and formulation of the stable and regular, though dynamic, relationships underlying two or more series of phenomena. We have stated previously that the fact of differences in geographic location, culture and time seems to preclude the possibility of developing any truly universal laws, or descriptions of relationships, among series of social phenomena which would be valid for all times, places and cultures. Therefore, our goal in developing a scientific sociology is necessarily limited in the description of relationships. Yet somehow, there seems to be a place for the sifting from sets of observed measures of relationships the irrelevant variations which particularize them as unique, in a search for meaningful relations, impermanent and varied with location though they be. This goal is so far short of those of the physical sciences that it may be misleading to use the term scientific in our field.

Without judging the issue for sociology, if nothing is constant, every situation is unique, then the past can't be a guide for the future and it is impossible to develop a predictive theory. There is no guarantee that application of scientific method to an arbitrary class of problems will uncover a predictive theory. It may be that no regularities or patterns exist for the class of problems. The fourth usage of the word science needs to be supplemented by a requirement that it achieve some success, perhaps yielding a descriptive or predictive theory.

The sign in front of the author's workplace reads, "Mathematical Sciences." Is mathematics a science? It is certainly systematized knowledge much concerned with structure, but then so is history. Does it employ the scientific method? Well, partly; hypothesis and deduction are the essence of mathematics and the search for counter examples is a mathematical counterpart of experimentation; but the question is not put *to nature*. Is mathematics about nature? In part. The hypotheses of most mathematics are suggested by some natural primitive concept, for it is difficult to think of interesting hypotheses concerning nonsense syllables and to check their consistency. However, it often happens that as a mathematical subject matures it tends to evolve away from the original concept which motivated it. Mathematics in its purest form is probably not natural science since it lacks the experimental aspect. Art is sometimes defined to be creative work displaying form, beauty and unusual perception. By this definition pure mathematics is clearly an art. On the other hand, applied mathematics, taking its hypotheses from real world concepts, is an attempt to describe nature. Applied mathematics, without regard to experimental verification, is in fact largely the "conditional truth" portion of science. If a body of applied mathematics has survived experimental test to become trustworthy belief then it is the essence of natural science.

13.2. Concepts of Statistics

Logical induction, distinguished from deduction, consists of reasoning from particular facts or individual cases to a general conclusion. Fisher considered statistical inference to be a solution to the old problem of how to infer conclusions inductively. Neymann and Pearson, and later Wald, aimed to base statistics on the "frequency of errors in judgment" resulting from "inductive behavior" Lehmann (1993, p. 1243) discusses the key role of "inductive inference versus inductive behavior" in the Fisher versus Neymann controversy.

We have presented Rubin's description of Bayesian statistical inference, and the personal degree of belief version of it, in Chapter 9. Rubin draws conclusions about unknowns on the basis of data. This too is a process of induction-inferring general statements from particular data.

Exploratory data analysis, as discussed by Velleman and Hoaglin (1992), aims to identify regularities and patterns in data. It thus emphasizes the process of discovery of generalities from particulars rather than the process of justification.

These four different philosophies of statistics are different paradigms for what constitutes rational inductive reasoning. Statistical inference, the subject, consists largely of models of inductive rationality.

13.3. Is Statistics a Science?

Box (1976, 1980) likens statistics to physics where one makes tentative assumptions about nature which are then checked by comparison with data. He suggests that progress in "the science of statistics" is to be realized in an analogous manner, the consulting practice of the statistician performing the function of comparison with data. The career of R. A. Fisher is presented in support. Wilks (1950, p. 1) expresses a similar view:

The test of the applicability of the mathematics in this field as in any other branch of applied mathematics consists in comparing the predictions as calculated from the mathematical model with what actually happens experimentally.

But Hogben (1957, p. 22) objects:

...what controlled experiment...would settle the dispute between Jeffreys and Fisher concerning the legitimacy of Bayes' postulate or the contest between Fisher and Neyman over test procedure.

In a footnote Wilks provides us with a reference: "For an example of such a comparison, see Ch. 5 of Bortkiewicz, *Die Iterationen,* Springer, Berlin, 1917." Bortkiewicz (1917) calls a repetition of a given event that takes place without interruption in a sequence of repeated trials, an *Iterationen;* the modern English term for an *Iterationen* is a "run." His book develops a number of probabilistic (mathematical) results concerning the theory of runs. His Chapter 5 exemplifies these mathematical results by comparison with several dichotomized empirical numerical series. For example, one series is the list of the sizes of the communities of the German Reich according to the census of 1910. Essentially, the mathematical model appears to fit these series rather well. Bortkiewicz may have hit upon a more-or-less permanent feature of demographic data, but stock market data might or might not display this feature.

Statisticians can and do make contributions to subject matter fields such as physics, and demography but statistical theory and methods proper, distinguished from their findings, are not like physics in that they are not about nature.

One does accumulate insight through statistical consulting experience with similar data sets. For example, miss distances of conventional military projectiles are well modeled as bi- or trivariate normal with occasional outliers, but this is a finding about ballistics rather than statistics. Applied statistics is natural science but the findings are about the subject matter field not statistical theory or method.

Statistical methods based on frequency probability can be checked by Monte Carlo experiment, comparing observed frequency with deduction from theory. Bayesian methods, at least those based on a degree of belief probability cannot be checked in this way, but experiments can be structured which observe the behavior of the believer. Such experiments check the consistency of conclusion with theory but not the ultimate rationality of theory or method.

If we accept that natural science is trustworthy conclusion concerning nature and that statistical inference, having to do with inductive rationality, is about mind not nature then statistics is not *natural* science; many of its most important concepts such as prior distributions and confidence intervals—the theories, not particular applications—are not about nature. Statistical theory does not employ "the scientific method." As Hogben tells us, statistical theory is not subject to experimental test.

To the extent that statistical theory is science, it augments and competes with **scientific method**, the art and science of natural science itself. Statistical theory helps with how to do natural science but it is not itself a natural science. Statistics is of course a science in the first weak sense that it is systematized knowledge. Statistics is also scientific in that it involves prediction and explanation. The utility of much of statistics is that it facilitates the prediction of future experiments on the basis of past experiments. Consistency of experimental outcome with theoretical prediction is the statistical criteria for success of a theory. Statistics attempts to guide and explain the conclusions which may be drawn from data.

14

Comparison of Evidential Theories

14.1. Evidential Models and Their Criteria

We start from the view that the purpose of statistical theory is to *explain* and *guide* what we choose to present as statistical evidence, i.e., what data and reasoning should cause us to adopt certain conclusions. We arrive at the position that there are kinds of statistical evidence, each of which lends a different explanatory insight and none of which is perfect. The explanatory models which we consider—along with their criteria—appear in Table 14.1.

14.2. Two Interpretations of the Mathematics of Testing Are Needed

Some authors recognize two traditions of statistical testing: (i) Neyman–Pearson–Wald hypothesis testing, which decides among behavioral responses to a performed experiment, based on the long-run error rates of the rules used for decision; and (ii) Fisherian significance testing—developed and exposited by Kempthorne and Folks (1971) and Cox and Hinkley (1974)—which considers p-values as an expression of experimental evidence. Much to Fisher's chagrin, Neyman denied the evidential interpretation (Neyman, 1950, p. 1; Fisher, 1956, p. 100; Johnston, 1986, p. 491). On the other hand, Fisher (1956, p. 91) tells us that the force or cogency of the evidence *is not* the infrequency with which the hypothesis is rejected in repeated sampling of any fixed population allowed by hypothesis. But he does not tell us what the force of the evidence is. We suggest that the force of the evidence is the degree of agreement between theoretical prediction and experimental outcome.

The mathematics used to formulate these very different tasks is similar; this is not surprising since both theories are attempts to explain and interpret the prior practice of computing significance level. As with all mathematical theories, if a theory of statistics is to have any practical meaning, then its basic terms must be interpreted and, on the other hand, intuitive statistical concepts can be modeled by

TABLE 14.1. Statistical models of induction and their explanations

Model	Criteria
Evidential meaning	Sufficiency and conditionality
Economic Bayes	Rational gambling
Propensity Bayes	Exchangeability
Likelihood inference	The "law" of likelihood
Hypothesis testing	Behavioral error rate
Show-me evidence	Agreement of prediction with experimental outcome

different theories. We think Neyman was right for the meanings which he gave to his mathematical concepts and Fisher was right for the different meaning which he gave to the same mathematical entities.

Lehmann (1993), suggests that evidential significance tests and Neyman–Pearson hypothesis tests might in fact be just different aspects of a single theory. But from the point-of-view of this exposition, their interpretations and explanations seem quite different. Some of the criticism of p-values is simply the result of trying to restructure their logic in an inappropriate behavioral context.

14.3. Evidential Meaning

Evidential meaning is less complete, as a model for induction, than the remaining models. The suggested conclusion here is the likelihood principle: that identical inferences, about a common unknown quantity τ, should be drawn from two different experiments resulting in proportional likelihood functions. But, as we have discussed in Chapter 8, no two real but different experiments can be formulated in terms of a common "true value," τ. Further, the evidential meaning of observing y as the outcome of experiment E will depend on the theoretical context, T, within which the experiment is placed. Data and theory do not speak for themselves; they have to be interpreted and the interpretation defended.

14.4. Comparison of Bayesian, Likelihood, and "Show-Me" Evidence

As an example of a theory T and its consequences, consider the Bayesian degree of belief logic—Savage (1962) and Berger and Berry (1988). From (9.1), identical Bayesian inferences are drawn from two different experiments resulting in proportional likelihood functions; Bayesian inference satisfies the likelihood principle.

Further, if $f(z|\theta_0)/f(z|\theta_1) = kf(y|\theta_0)/f(y|\theta_1)$ then the posterior degree of belief in θ_0 relative to θ_1 upon observing z, is k times stronger than upon observing y. Hence $r(y) = f(y|\theta_0)/f(y|\theta_1)$ is a Bayesian measure, on a ratio scale, of the evidential strength with which the observation y supports θ_0 over θ_1.

From this Bayesian starting point is developed the theory of likelihood inference, a data analysis method similar in spirit to, but different in result from, hypothesis testing—see Royall (1997). Royall and DeGroot (1975, p. 380) assume that the Bayesian evidential ratio scale interpretation continues to apply in the absence of the Bayesian context. Royall defines statistical evidence in terms of the likelihood ratio—the "law" of likelihood. On page 13 he supports his definition with an analogy from the physics of heat; in fact his theory of likelihood inference rests on this pillar, since within the Bayesian model, Bayesian—not likelihood—inference is appropriate. It is a judgment call. The reader may wish to examine Royall's analogy to judge whether it is strong enough to support an entire theory of inference. We judge that, without the Bayesian model, there is little reason why the likelihood ratio should measure strength of evidence on a ratio scale. If it isn't meaningful to talk about probabilities of hypotheses, then it is even less meaningful to talk about k fold increases in ratios of such probabilities.

For convenience, we repeat Assumption i′ of show-me evidence (Section 11.2). For the two densities f_η and f_κ, the likelihood ratio $r(y) = f_\eta(y)/f_\kappa(y)$ orders the sample points of S according to their agreement with η (relative to κ).

There is substantial agreement, among statistical theories, on the ordinal significance of the likelihood ratio. The various theories differ primarily on reasons and interpretations; the Neyman–Pearson lemma is the foremost example. As we have observed, the Bayesian theory agrees on the ordinal significance of the likelihood ratio but goes farther.

Without the Bayesian structure, and for simple hypotheses H and K, the sufficiency principle still tells us that if $f_H(z)/f_K(z) = f_H(y)/f_K(y)$, then y and z have equal evidential meaning (Cox and Hinkley, 1974, pp. 20, 92). It is then highly intuitive that smaller likelihood ratios indicate stronger evidence against H. This intuition can be supported by theory—Theorem 11.1, for example.

We may contrast three theories as follows. The Bayesian model introduces postulates which imply that the likelihood ratio measures strength of Bayesian evidence on a ratio scale. Likelihood inference defines strength of *all* statistical evidence to be the likelihood ratio on a ratio scale. Show-me inference takes the likelihood ratio as a measure of strength of show me evidence but only on an ordinal scale. The three theories provide different grounds for conclusion under different assumptions about which people may reasonably disagree.

Contrasted with hypothesis testing and economic Bayesianism, the spirit of show-me evidence is that, while long run error rates of statistical procedures and appropriate personal betting odds lend some insight, people are convinced primarily by agreement of theoretical prediction with experimental outcome. We think *p*-values have something to contribute but we also like Bayesian inference; there are kinds of evidence. The greater assumed structure of the Bayes approach facilitates stronger conclusions and consideration of more problems.

But sometimes it may be difficult to visualize Y and Θ as having a joint probability distribution, in which case Bayes' theorem of probability does not apply and Bayesian inference is not indicated. It isn't so much *which* distribution should be specified but rather whether the situation at hand conforms to any probability law. It isn't clear that all quantities may be treated as random variables nor that all inference situations may be treated as chance experiments. Of course, we may make the prior probability assumption as an unsupported leap of faith. But then we are attributing an aura of scientific certainty to a purely speculative situation. At this writing, one of the most complete and successful explanations of Bayesian inference interprets probability as personal degree of belief through consideration of "economic man." But one encounters substantial difficulties in extending that economic argument to the conclusions of a group-the concern of science. See our Section 9.4. Alternatively, we may explain Bayesian statistics as a frequency theory applicable to exchangeable r.v.s . This theory is not helpful in choosing a prior and does not work for *i.i.d.* random variables.

Berger and Berry (1988) and Howson and Urbach (1989) suggest the normative approach that scientists should adopt the Bayesian model of inference, updating personal belief as a consequence of new data to answer the question: "What now should I think?" But personal evaluation is not the evidential criterion of science; science is social not personal. Agreement of experimental outcome with theoretical prediction is the scientific criterion for the success of a theory.

The claim that scientists should alter their traditional methods of inference, to conform with the Bayesian paradigm, deserves consideration. But only by happy accident can the Bayesian model be a description of how science *does* work, for that would be an anachronism; the scientific spirit predates Bayes by hundreds of years.

Marked differences between Bayesian theory and scientific practice are, the treatment of failed hypotheses and the possibility of innovation. When an idea is contradicted by experiment, science does not simply adjust background (prior) knowledge in the light of the data. All Bayesian prior opinion is updated using Bayes' theorem whereas science simply throws failed hypotheses in the trash. That is the meaning of the cliché "back to the drawing board."

Bayesian updating results in the second difference. Scientific knowledge typically grows as new conclusions are arrived at as a consequence of observing new phenomena. In contrast, all Bayesian posterior ideas must be among those having positive prior probability, there is no possibility of introducing things not previously considered.

In another context, the importance of not working with a closed set of possibilities is remarked on by Ackoff (1979):

Creative solutions to problems are not ones obtained by selecting the best from among a well- or widely recognized set of alternatives, but rather by finding or producing a new alternative. Such an alternative is frequently so superior to any of those previously perceived that formal evaluation is not required.

And again by Dyson (1988, p. 196)—

The difficulty in imagining the future comes from the fact that the important changes are not quantitative. The important changes are qualitative, not bigger and better rockets but new styles of architecture, new rules by which the game of exploration is played.

The distinction "Bayesian" is not just whether the parameters of a sampling distribution are fixed or random. A parameter is a quantity which varies according to circumstances but is considered constant for a given circumstance. A fixed parameter is considered to be an unspecified constant. A random parameter is a quantity considered to be a random variable in the sense of probability theory.

Eisenhart (1947) explains that the parameters of non Bayesian models may either be fixed or random. For example, the astronomer Airy (1861) considers a series of observations of the same quantity over several nights. His model for the j-th observation on the k-th night is $y_{kj} = \mu + a_k + e_{kj}$ where μ is the true, and hence fixed but unknown, value of the observation being made while a_k is the random k-th night effect (or parameter) caused by "atmospheric and personal circumstances"; e_{kj} is the error of the kj-th observation about its conditional mean $\mu + a_k$. Rather than the randomness of the parameters, the important dichotomy in inferential philosophy is the old one of mind versus matter (see Durant, 1953).

14.5. An Example—Probability Evidence and the Law

The following question arises in several legal contexts: An accused is capable of leaving some incriminating trace which was left by some perpetrator. In this situation, what are the implications of "probability evidence" regarding the guilt of the accused? Statistical instruments which have been considered are population proportions, p-values and posterior odds (or probabilities) of guilt.

The general legal situation is that "...most courts admit decent evidence of population proportions" (Kaye, 1987, p. 161), but as was established in the eye-witnesses testimony case of people vs. Collins (our Section 4.1)without some empirical justification of prior probabilities, hypothetical probabilistic and statistical arguments are inadmissible. There is a difficulty in choosing the relevant population. For example, in a case like Collins should one consider all the individuals in Los Angeles, or California or the nation? Or should the population consist of couples? There is no "true" value. But to proceed, assume this difficulty resolved.

The population proportion, β, may be given two alternative interpretations. First, if η denotes the hypothesis that the defendant is randomly chosen from the population and I is the event that the defendant is capable of leaving the incriminating trace then $\beta = pr(I \mid \eta)$.

Second, β turns out to be the show-me evidence, of Chapter 11, for guilt relative to the "background knowledge" of either guilt or random choice of defendant from the population. To see this let κ be the hypothesis that the defendant is the perpetrator and let Y indicate whether the defendant is capable of leaving the incriminating trace or not, that is $Y = 1$ or 0 according as I is or is not the case. We have

$$f(y|\kappa) = \begin{cases} 1, y = 1 \\ 0, y = 0 \end{cases}$$

$$f(y|\eta) = \begin{cases} \beta, y = 1 \\ 1 - \beta, y = 0 \end{cases}$$

and

$$r(y) = \frac{f(y|\eta)}{f(y|\kappa)} = \begin{cases} \beta, y = 1 \\ \infty, y = 0. \end{cases}$$

Now $p(1; \eta, \kappa) = pr_\eta [r(Y) \le r(1)] = pr_\eta(Y = 1) = \beta$, so

$$p(y; \eta, \kappa) = \begin{cases} \beta, y = 1 \\ 1, y = 0 \end{cases}$$

$$p(y; \kappa, \eta) = \begin{cases} 1, y = 1 \\ 0, y = 0 \end{cases}$$

And the bivariate evidence for κ—guilt of the defendant—is

$$ev(y) = \begin{cases} (\beta, 1), y = 1 \\ (1, 0), y = 0 \end{cases}$$

which depends only on the size of β. If the defendant cannot have left the incriminating trace, then of course he/she is innocent, but if $Y = 1$ and β is small then there is strong evidence of guilt relative to the hypothesis of random choice.

Rather than $\beta = pr(I |\eta)$, we would really like $pr(\kappa |I)$. Bayes' theorem gives a formula for this probability; an equivalent but neater expression is obtained in terms of odds:

$$odds(\kappa |I) = \frac{pr(I |\kappa)}{pr(I |not \kappa)} odds(\kappa). \tag{14.1}$$

Kaye (1987, p. 171) casts a no vote on this use of Bayes' theorem:

I would not be the first to insist that having an expert give an opinion as to the "probability of guilt" or the probability that the defendant is the one who left trace evidence is inadvisable. At best, a forensic scientist can testify to the probability of certain observations (like matching blood groups) assuming that the defendant left the incriminating evidence, or assuming that someone else did. But these scientists have no special claim to being able to figure out the inverse probability that a defendant is or is not guilty given the evidence. Such statements require knowledge of the prior probability of guilt. How can experts know this?

According to Mueller and Kirkpatrick (1999, pps. 960, 970) formula (14.1) is routinely applied to paternity cases in the form

$$odds(\kappa \mid I) = \frac{pr(I \mid \kappa)}{pr(I \mid \eta)} \cdot 1 = \beta^{-1}. \tag{14.2}$$

Mueller and Kirkpatrick (1999, p. 968) explain that the prior odds are taken to be even "(meaning one chance that a defendant is the father as against one chance that he is not). This number is chosen because it is supposedly neutral between the positions of the plaintiff (claiming paternity) and the position of the defendant (resisting)." This is a new and interesting explanation or interpretation of the Bayesian logic. The legal attitude toward probability evidence varies with the context, p-values receive a lukewarm reception because of the danger of misinterpreting $pr(I \mid \eta)$ as $pr(I \mid not \, \kappa)$ or worse $pr(not \, \kappa \mid I)$ (Kaye, 1987, p. 168). And yet the formula (14.2) which makes the same misinterpretation—substituting $pr(I \mid \eta)$ for $pr(I \mid not \, \kappa)$—is routinely applied to paternity cases. It is unreasonable to assume that, if the defendant is not the father, then the father must have been randomly chosen.

Formula (14.2) makes two assumptions: (i) $\eta = not \, \kappa$ and (ii) $pr(\kappa) = pr(not \, \kappa)$. So $pr(\eta) = pr(not \kappa). = 1/2$—a relatively large value. But η seems a licentious and unlikely a priori mechanism for choosing a father; so $pr(\eta)$ should be small—an apparent internal inconsistency.

Mueller and Kirkpatrick (1999, p. 961) summarize the legal attitude toward statistical evidence as follows:

For various kinds of trace evidence, . . . including hair samples and DNA, courts generally admit at least statistics reflecting the scarcity of the sample, and occasionally similar statistics relating to the probability of other natural phenomena. Indeed, in some criminal cases where proof that the defendant is the father of a child would itself constitute proof of intercourse, hence guilt of the charged crime, some courts even admit testimony on the . . . probability of paternity. In contrast to Collins and the blue bus cases, where the facts are personal, social, and economic, most of these cases involve physical and medical facts. Here probability data are less likely to reflect human choice or volition, where we are less comfortable resting specific decisions about particular people on classwide generalizations. These differences may account for the difference in judicial treatment.

Appendix A

Deductive Logic

A.1. The Propositional Calculus

Logic, the study of correct reasoning, is traditionally divided into two parts: deduction, reasoning from general principles without introducing falsehood, and induction, inferring generalities from particular instances.

Our immediate concern is with deduction. The usual modern approach to deduction, and particularly to mathematics, is through the propositional calculus. A **proposition**, or statement, is the meaning of a declarative sentence. Thus, questions or exclamations are not propositions, and the same declaration expressed in two different ways, say French and English, is the same proposition.

The propositional calculus is about truth and falsity. Since Aristotle (384–322 B.C.) announced the "law of the excluded middle," that each proposition is either true or false, it has been fundamental to the mainstream of Western logical thought that these are the only two possibilities. Within the calculus the meaning of truth is left undefined, but when the calculus is applied to some concept, truth is popularly thought to correspond to the way things are.

The propositional calculus hypothesizes a universal set, $U = \{x,y,z,\ldots\}$, of logical possibilities and considers the truth or falsity of various propositions, p, q, etc., relative to U. Different listings of the logical possibilities will be appropriate for different purposes.

Example A.1 A red and green ball are each placed in one of three cells. The possibilities appear in Table A.1.

Example A.2 If the balls of Example A.1 are regarded as indistinguishable, as they might appear to a color-blind person, then we obtain a different set of logical possibilities, which are displayed in Table A.2.

Denote the **truth value** of p for logical possibility x by $|p,x|$ and write $|p,x| = 1$ or 0 according as p is true or false for x. The **truth set** P of proposition p is $P = \{x: |p,x| = 1\}$. Conversely, truth values may be obtained from the truth set. In fact, $|p,x|$ equals 1 or 0 according as possibility x is or is not in the set P.

133

TABLE A.1. Cell placement possibilities for two balls in three cells

Possibility\color	Red	Green
y_1	1	1
y_2	1	2
y_3	1	3
y_4	2	1
y_5	2	2
y_6	2	3
y_7	3	1
y_8	3	2
y_9	3	3

A proposition that is true for all logical possibilities (has U for truth set) is called a **tautology**; such statements are true, regardless of circumstances, as a consequence of their logical structure. On the other hand, a proposition that is false for all logical possibilities (has empty truth set) is called a **contradiction**. Those propositions that are true for some logical possibilities but false for others are called **contingent**.

Simple propositions may be combined to form compound propositions. The **negation** $\sim p$, read not p, is the proposition that states exactly the opposite of p. The **disjunction** $p \vee q$, read p or q, is clearly true when at least one of its components is true. The **conjunction** $p \wedge q$, read p and q, will be true if both components are true.

We may formalize the above observations explicitly as follows: for all $x \; \varepsilon \; U$,

$$|\sim p, x| = 1 - |p, x| \tag{A.1}$$

$$|p \vee q, x| = \max(|p, x|, |q, x|), \tag{A.2}$$

$$|p \wedge q, x| = \min(|p, x|, |q, x|). \tag{A.3}$$

For brevity, here and in similar contexts, we may suppress the roles of possibility and universal set. We then write, for example,

$$|\sim p| = 1 - |p|, \tag{A.1'}$$

$$|p \vee q| = \max(|p|, |q|), \tag{A.2'}$$

TABLE A.2. Number of indistinguishable balls placed in the three cells

Possibility\Cell	1	2	3
x_1	2	0	0
x_2	0	2	0
x_3	0	0	2
x_4	1	1	0
x_5	1	0	1
x_6	0	1	1

TABLE A.3. Some famous "laws" of classical logic

	Law	Linguistic statement	Symbolic statement
i.	Double negation		$\sim(\sim p) = p$
ii.	contradiction	no proposition is simul-	$\vert \sim p \wedge p \vert = 0$
		taneously true and false	
iii.	deMorgan		$p \wedge q = \sim(\sim p \vee \sim q)$
iv.	excluded middle	each proposition is	$\vert p \wedge \sim p \vert = 1$
		either true or false	

and

$$\vert p \wedge q \vert = \min(\vert p \vert, \vert q \vert). \tag{A.3$'$}$$

We say that p and q are **equivalent** and write $p = q$ if $P = Q$ or if $\vert p,x \vert = \vert q,x \vert$ for all $x \ \varepsilon \ U$. The truth sets of $\sim p$, $p \vee q$, and $p \wedge q$ are $\bar{P}$, $P \cup Q$ and $P \cap Q$. $\bar{P}$ is the complement of P, and $\cup$ and $\cap$ denote union and intersection, respectively.

For illustration and completeness several famous rules of classical logic are given in Table A.3. Their proofs from the present point of view are as follows:

i. $\vert \sim(<nl>\sim p) \vert = 1 - \vert \sim p \vert = 1 - (1 - \vert p \vert) = \vert p \vert.$

ii. $\vert \sim p \wedge p \vert = \min(1 - \vert p \vert, \vert p \vert) = 0,$

iii. $\vert \sim(\sim p \vee \sim q) \vert = 1 - \vert \sim p \vee \sim q \vert = 1 - \max(1 - \vert p \vert, 1 - \vert q \vert) = \min(\vert p \vert, \vert q \vert) = \vert p \wedge q \vert.$

iv. $\vert p \vee \sim p \vert = \vert \sim(\sim p \wedge p) \vert = 1 - \vert \sim p \wedge p \vert = 1.$

A.2. Truth Tables

A useful alternative way of defining truth values of compound statements is the device of **truth tables**, which amounts to one particular choice of universal set of logical possibilities. All possible true and false combinations of component statements are enumerated and displayed in matrix array. For example, the truth tables corresponding to (A.1), (A.2), and (A.3) appear as Tables A.4 (a) and (b).

TABLE A.4(a) Truth table defining negation

	(1.1)
p	$\sim p$
1	0
0	1

TABLE A.4(b) Truth tables defining disjunction and conjunction

p	q	(1.2) $p \lor q$	(1.3) $p \land q$
1	1	1	1
1	0	1	0
0	1	1	0
0	0	0	0

TABLE A.5 A truth table for indistinguishable balls

Possibility	p	q	r	$\sim p$	$p \lor q$	$p \land q$
x_1	0	0	1	1	0	0
x_2	0	1	1	1	1	0
x_3	0	0	1	1	0	0
x_4	1	0	1	0	1	0
x_5	1	0	1	0	1	0
x_6	0	0	1	1	0	0

Example A.2 (cont.) We employ Example A.2 and Table A.2 to illustrate the concepts presented thus far. Consider the propositions p = (one ball is placed in the first cell), q = (two balls are placed in the second cell), and r = (the sum of the numbers of balls contained in all cells is two). Truth values of various propositions appear in Table A.5.

We have that $U = \{x_1, \ldots, x_6\}$ and $R = U$; hence, r is a tautology. $P = \{x_4, x_5\}$ and $Q = \{x_2\}$. The truth set of $p \lor q$ is $\{x_2, x_4, x_5\} = P \cup Q$. The proposition $p \land q$ is a contradiction since its truth set is null. Note that many true and false combinations are not logically possible; for example, $|r| = 0$ is missing.

A.3. Deductive Arguments and Conditional Statements

Becoming more technical, deduction is the study of valid arguments. A **deductive argument** is an assertion that one statement, the conclusion, necessarily follows from other statements, the premises. The word "necessarily" is important here; it means independent of the true or false state of the world. A deductive argument is said to be **valid** if the conclusion is true for all logical possibilities for which all premises are true. An invalid argument is called a **fallacy.**

An alternative way of discussing arguments and their validity is in terms of implication. If q is a conclusion and p is the conjunction of the premises, then we say p **implies** q, or write symbolically $p \Rightarrow q$, to mean that q necessarily follows from p is a valid argument (i.e., whenever $x \in U$ and $|p,x| = 1$, then $|q,x| = 1$). Implication, $p \Rightarrow q$, is variously expressed as p is a sufficient condition for q, and q is a necessary condition for p. The essence of implication and the essence of what constitutes a valid deductive argument is that truth is preserved.

Directly from the definition we see that $p \Rightarrow q$ is the requirement that $P \subset Q$. Hence, $p \Rightarrow q$ and $q \Rightarrow p$ is the requirement that $P = Q$ or $p = q$.

An important relation between arguments and propositions is the following.

Theorem A.1 $p \Rightarrow q$ is equivalent to requiring $\sim p \vee q$ to be a tautology.

Proof
Consider $x \; \varepsilon \; U$. If $\sim p \vee q$ is a tautology, then $1 = |\sim p \vee q, x| = \max\{1 - |p,x|, |q,x|\}$, and if $|p,x| = 1$, then $|q,x| = 1$. Conversely, if $p \Rightarrow q$, then either $|p,x| = 0$ or 1; in the latter case $|q,x| = 1$. Hence, in either case $|\sim p \vee q, x| = \max\{1 - |p,x|, |q,x|\} = 1$.

Thus, the compound proposition $\sim p \vee q$ is particularly important in analyzing the validity of arguments, and a special logical connective,"$\rightarrow$ ", called the **conditional**, is introduced: $p \rightarrow q$ is written to mean $\sim p \vee q$. The compound proposition $p \rightarrow q$ may be read "if p then q." In discussing the conditional, p is called the **antecedent** and q the **consequent**.

Theorem A.1 states that $p \Rightarrow q$ and $p \rightarrow q$ are related; but there is a difference. The difference is that $p \rightarrow q$ is a proposition that is true or false depending on x, while $p \Rightarrow q$ is an argument which is valid if $\sim p \vee q$ is true for all $x \; \varepsilon \; U$.

Table A.6 is the truth table of the conditional. Two comments on Table A.6, sometimes called paradoxes of the conditional, are in order. First, $p \rightarrow q$ is not to be given a causal interpretation. For example, if p is the proposition that salt is white and q that cows eat grass, then we have possibility x_a and $|p \rightarrow q, x_a| = 1$, but the color of salt does not affect the diet of cows. It is simply the case that p and q are both true and, hence, truth is preserved. Second, for x_c and x_d, $p \rightarrow q$ is true by default. The way to look at this is as follows. Suppose $|p,x| = 0$. Does $p \rightarrow q$ fail to preserve truth? The answer is no, regardless of the truth value of q; there is no truth of p to be preserved. Hence, $p \rightarrow q$ is not false and must, therefore, be true as a consequence of the law of the excluded middle.

A sufficient condition for an argument to be valid is that, in a truth table, the conclusion is true whenever all premises are true.

Example A.3 Affirming the antecedent
The premises of this argument are p and $p \rightarrow q$; the conclusion is q. Observe from Table A.6 that x_a is the only possibility where p and $p \rightarrow q$ are both true, and, there, q is also true. Therefore, affirming the antecedent is a valid argument.

TABLE A.6 Truth table of the conditional

Possibility	p	q	$p \rightarrow q$
x_a	1	1	1
x_b	1	0	0
x_c	0	1	1
x_d	0	0	1

Example A.4 Denying the consequent

Observe from Table A.6 that x_d is the only case where the premises $p \rightarrow q$ and $\sim q$ are both true, and, there, the conclusion $\sim p$ is true. Denying the consequent is, therefore, a valid argument.

However, the argument $(p \rightarrow q) \wedge q \Rightarrow p$, called affirming the consequent, is, in general, a fallacy since from Table A.6 $| p \rightarrow q, x_c | = 1$ and $|q, x_c| = 1$ but $|p, x_c| = 0$. However, if x_c is not a logical possibility, then the argument becomes valid. Similarly, $(p \rightarrow q) \wedge \sim p \Rightarrow \sim q$, denying the antecedent, is, in general, a fallacy.

Some invalid arguments are difficult to formulate and expose in terms of the propositional calculus; they must be recognized by their specific content.

Appeal to ignorance, that if we do not know q, then q must be false, is such an informal fallacy. However, another argument, which is easily confused with appeal to ignorance, is valid. Consider the following paraphrase of an argument that appeared in a United States government pamphlet.

AIDS cannot be transmitted by mosquito. For AIDS is now the most thoroughly researched communicable disease, and every case thus far has been traced to a source other than mosquitos. This is in spite of the fact that AIDS is most prevalent in areas of the world, notably central Africa, where mosquitos are common. If mosquitos could transmit AIDS, then at least one such case would have shown up by now.

Let $p = $ (AIDS can be transmitted by mosquito) and $q = $ (we know p). The AIDS argument is that if p were true, then we would know it, and we do not know p so p is not true. Symbolically, this is $(p \rightarrow q) \wedge \sim q \Rightarrow \sim p$, a valid argument.

The scientific method of theory checking is to compare predictions deduced from a theoretical model with observations on nature. A deductive analysis of how this works is the following. Any investigation will be in the context of background knowledge, m, used for constructing and interpreting the question to be put to nature. At issue is a hypothesis h. Using m and h, a prediction is deduced that in one particular experimental instance nature will answer p. The particular instance is carried out and $m \wedge p$ is observed to be true or false.

Table A.7 is used to interpret the results. Since $m \wedge h \Rightarrow p$, or $\sim(m \wedge h) \vee p$ is a tautology, the situation $|m| = |h| = 1$, $|p| = 0$ cannot be a logical possibility

TABLE A.7 Deductive analysis of scientific theory checking

m	h	p	$m \wedge h$	$\sim(m \wedge h) \vee p$	$m \wedge p$	$m \wedge \sim p$
1	1	1	1	1	1	0
+	+	0	+	0	0	+
1	0	1	0	1	1	0
1	0	0	0	1	0	1
0	1	1	0	1	0	0
0	1	0	0	1	0	0
0	0	1	0	1	0	0
0	0	0	0	1	0	0

as indicated by "marking out" the second line of Table A.7. If $|m \wedge p| = 0$ we cannot conclude $|h| = 0$, only that $|m \wedge h| = 0$; or in the words of Ayer (1982, p. 11) "...the propositions of scientific theory face the verdict of experience, not individually but as a whole." On the other hand, if $|m \wedge p| = 1$ then h as well as $m \wedge h$ may be either true or false. Or as Karl Popper puts it, the falsity but not the truth of generalizations can be deduced from appropriate particulars. If $|m \wedge p| = |m \wedge \sim p| = 0$ then the appropriate conclusion, is $|m| = 0$, to question background knowledge.

A.4. Critique of Classical Logic

It would be difficult to overestimate the importance of classical two-valued logic. For "the man in the street" it is a dull subject and yet many of the great achievements of the human mind have been made to depend on classical logic; two obvious examples are mathematics and computers. However, it is not at all difficult to find instances where classical reasoning is clearly inappropriate. A major problem with the deductive analysis of Table A.7, discussed in Chapter 3 and later, is that models and hypotheses are not true or false, only more or less accurate for some purpose.

Now turning to another difficulty with deduction: we understand that Aristotle himself observed that future contingent statements, such as the outcome of tomorrow's sea battle, cause difficulties for his law of the excluded middle. "We will win the battle tomorrow," stated the day before seems to be indeterminate rather than true or false. "We did win the battle yesterday," reported the day after, makes precisely the same claim about the facts, but from a different temporal perspective. The claim changes from indeterminate to either true or false with the passage of time. The calculus of propositions provides for neither indeterminate nor changed truth values. This still unresolved issue seems to have important implications for statistics. Aristotle's sea battle as predicted the day before or reported the day after, seems to be like the outcome of a sampling experiment before and after its performance.

Two friends disagree on the role of education in a democracy. One believes that all citizens must receive higher education in order to vote competently, the other that only capable students should receive higher education because the presence of the slower student dilutes the quality of the whole, to the detriment of the state. Each reasons, from the law of the excluded middle, that since he is right and their positions are incompatible, that the other is wrong. Must one of them be right and the other wrong, though perhaps we do not know which? Or, are each of their positions only partly true but a little false?

Even in mathematics, that stronghold of classical logic, there is a dissenting minority of "intuitionists," led by Brouwer, who reject the law of the excluded middle. They employ a logic in which $p \vee \sim p$ is not a tautology.

Before truth can correspond to "the way things are," things must be just one way. The truth of a proposition must be invariant to changes of conditions not covered in the proposition. Truth must be invariant with respect to time perspective, the same after the sea battle as before. Truth must be the same at different locations; a chemical experiment must give the same result at different laboratories. Changing the observer must not change truth value, my truth must be your truth. Either truth value must be unchanged by altered conditions of observation, or the proposition must specify the effect of the alteration.

In summary, while classical logic has proved to be very useful, invariance qualifications are needed before the "universal" laws of logic are valid. Furthermore, the precise nature of these qualifications is not at all clear, and an adequate substitute for classical logic has not presented itself.

A.5. Many-Valued Logic

As noted in the previous section, there are compelling reasons for considering logical systems that incorporate truth values other than traditional truth and falsity. Listen to what motorcycle maintenance has to say about two-and three-valued logic.

Because we're unaccustomed to it, we don't usually see that there's a third possible logical term equal to yes and no which is capable of expanding our understanding in an unrecognized direction. We don't even have a term for it, so I'll have to use the Japanese mu...(Mu) states that the context of the question is such that a yes or no answer is in error and should not be given.

The dualistic mind tends to think of mu occurrences in nature as a kind of contextual cheating, or irrelevance, but mu is found throughout all scientific investigation, and nature doesn't cheat, and nature's answers are never irrelevant. It is a great mistake, a kind of dishonesty, to sweep nature's mu answers under the carpet. Recognition and valuation of these would do a lot to bring logical theory closer to experimental practice. Every laboratory scientist knows that very often his experimental results provide mu answers to the yes-no questions the experiments were designed for. In these cases he considers the experiment poorly designed, chides himself for stupidity and at best considers the "wasted" experiment which has provided the mu answer to be a kind of wheel-spinning which might help prevent mistakes in the design of future yes–no experiments.

This low evaluation of the experiment which provided the mu answer isn't justified. The mu answer is an important one. It's told the scientist that the context of his question is too small for nature's answer and that he must enlarge the context of the question. That is a very important answer! His understanding of nature is tremendously improved by it, which was the purpose of the experiment in the first place. A very strong case can be made for the statement that science grows by its mu answers more than by its yes or no answers. Yes or no confirms or denies a hypothesis. Mu says the answer is beyond the hypothesis. Mu is the "phenomenon" that inspires scientific inquiry in the first place! There's nothing mysterious or esoteric about it. It's just that our culture has warped us to make a low value judgement of it. In motorcycle maintenance the mu answer given by the machine to many of the diagnostic questions put to it is a major cause of gumption loss. It shouldn't be! When your answer to a test is indeterminate, it means one of two things: that your test procedures aren't doing what you think they are or that your understanding of the context of the question needs to be enlarged. Check your tests and restudy the question. Don't throw away those mu answers! They're more vital. They're the ones you grow on!

Persig (1984, p. 288, et seq.)

TABLE A.8 Lukasiewicz's three-valued logic

~p	p\q	p∧q 1 I 0	p∨q 1 I 0	(p→q) = (~p∨q) 1 I 0
0	1	1 I 0	1 1 1	1 I 0
I	I	1 I 0	1 I I	1 I I
1	0	0 0 0	1 I 0	1 1 1

An observation on Persig is that, while he makes an interesting point, it does not establish the necessity for multiple-valued logic. In Table A.7, we have analyzed the situation without introducing mu. Let m be the background knowledge or model used for constructing and interpreting the diagnostic question put to the motorcycle and let p be a theoretical deduction of the cycle's answer. If $|m\wedge p| = |m\wedge \sim p| = 0$ then $|m| = 0$.

The best known generalization of the classical propositional calculus is Table A.8, the three-valued logic of Lukasiewicz. If I is thought of as being intermediate between 0 and 1 in truth value, then equations (1.2) and (1.3) hold here as well. There are numerous other systems of many-valued logic besides the three-valued logic of Lukasiewicz. Most of these take the formulae (A.1), (A.2), and (A.3) as definitions of negation, disjunction, and conjunction. The following guidelines have been widely adopted: (a) the truth value of the negation of a statement is its "mirror image" in truthfulness, (b) the truth value of a disjunction is the truest of the truth values of its components, and (c) the truth value of a conjunction is the falsest of the truth values of its components. In particular, there is an infinite version of Lukasiewicz logic, which contemplates all possible truth values between 0, denoting absolute falsity, and 1, denoting absolute truth. But note that the third truth value in the problem of future contingency, mentioned in Section A.4, and Persig's mu may *not* be intermediate between true and false. That third value may represent indeterminate, inapplicable, undefined, confused, muddled, inappropriate, or out of context.

In the propositional calculus a proposition is called a tautology if it is true for all logical possibilities. This concept is readily extended to many valued systems by **designating** certain of its truth values as being nearly true. A proposition, p, is then called a **tautology** of the many valued system if $|p,x|$ is designated for all logical possibilities $x \in U$. Similarly, we may antidesignate certain truth values as being nearly false and define p to be a many valued **contradiction** if $|p,x|$ is antidesignated for all $x \in U$. Observe that the negation of a many valued tautology will be a contradiction and, vice versa, (as in the propositional calculus) if the truth function for negation has the "mirror image" property of taking designated truth values into antidesignated ones and vice versa.

In the early 1900's, when logic was being systematized and studied abstractly, truth functionality was thought to be an essential feature. A **connective** ϕ is **truth functional** if there is a function F_ϕ such that $|\phi(p,q),x| = F_\phi(|p,x|,|q,x|)$. A **logic** is **truth functional** if each of its connectives are. The propositional calculus and Lukasiewicz's three-valued system are both truth functional with $F_\sim(1) = 0$,

$F_\sim(0) = 1$, $F_\sim(I) = I$, $F_\vee(u,v) = \max(u,v)$ and $F_\wedge(u,v) = \min(u,v)$. It is truth functionality that makes truth table methods possible; and, of course, a logic will be truth functional if each of its connectives may be defined in terms of a truth table.

Notes on the Literature

Most of the first four sections are standard and can be found in any book on logic. An exception is the explicit introduction of the possibility space U. Many writers handle this aspect linguistically. For example, they describe a tautology as a statement that is true no matter what the facts. We prefer the concrete possibility space language. A more extensive treatment close to the above is Kemeny et al. (1958). For more detail concerning many-valued logic, the reader may consult Rescher (1969).

References

Ackoff, R.L. (1979) The Future of Operations Research is Past. *The Journal of the Operational Research Society*, **30**(2), 93–104.

Aczél, J. (1966) *Lectures on Functional Equations and Their Application's*. New York: Academic Press.

Airy, G.B. (1861) *On the Algebraic and Numerical Theory of Errors of Observations and the Combination of Observations*. London: Macmillan and Co.

Anderson, T.W. (1958) *An Introduction to Multivariate Statistical Analysis*. New York: Wiley.

ASTM (1999) *Annual Book of ASTM Standards*. Philadelphia.

Ayer, A.J. (1982) *Philosophy in the Twentieth Century*. London: Unwin.

Barnard, G.A. (1980) Discussion of Box, G.P. Sampling and Bayes' Inference in Scientific Modeling and Robustness. *Journal of the Royal Statistical Society, Series A*, **143**, 404–406.

Barnett, V. (1982) *Comparative Statistical Inference*, 2nd ed. New York: Wiley.

Berger, J.O. and Wolpert, R.L. (1984) *The Likelihood Principle*. Hayward, CA: Institute of Mathematical Statistics.

Berger, J.O. and Berry, D.A. (1988) Statistical Analysis and the Illusion of Objectivity. *American Scientist,* **76**, 159–165.

Berger, J.O. and Sellke, T. (1987) Testing a Point Null Hypothesis: The Irreconcilability of *P*-Values and Evidence. *Journal of American Statistical Association*, **82**, 112–139.

Berkson, J. (1942) Tests of Significance Considered as Evidence. *Journal of the American Statistical Association*, **37**, 325–335.

Birnbaum, A. (1962) On the Foundations of Statistical Inference (with Discussion). *Journal of the American Statistical Association*, **53**, 259–326.

Blumenthal, L.M. (1980) *A Modern View of Geometry*. New York: Dover.

Bortkiewicz (1917) *Die Iterationen*. Berlin: Springer.

Box, G.E.P. (1976) Science and Statistics. *Journal of the American Statistical Association,* **71**, 791–799.

Box, G.E.P. (1980) Sampling and Bayes' Inference in Scientific Modeling and Robustness. *Journal of the Royal Statistical Society. Series A*, **143**, 383–430.

Carson, D.A. (2003) The Dangers and Delights of Postmodernism. *Modern Reformation Magazine*, **12.4**, 1–7.

Casella, G. and Berger, R. (1987) Reconciling Bayesian and Frequentist Evidence in the One-Sided Testing Problem. *Journal of American Statistical Association*, **82**, 106–111.

Chalmers, A.F. (1982) *What is this thing called Science?* St.Lucia: University of Queensland Press.

Cox, D.R. and Hinkley, D.V. (1974) *Theoretical Statistics.* London: Chapman and Hall.

Cramér, H. (1946) *Mathematical Methods of Statistics.* Princeton: Princeton University Press.

Darwin, C. (1859) *The Origin of Species.* London: Murray.

Davies, P.L and Kovak, A. (2001) Local Extremes, Runs, Strings and Multiresolution. *The Annals of Statistics*, **29**, 1–47.

DeGroot, M.H. (1974) Reaching a Consensus. *Journal of the American Statistical Association,* **69**, 118–121.

DeGroot, M.H. (1975) *Probability and Statistics*, 2nd ed. Reading: Addison-Wesley.

Deming, W.E. (1950) *Some Theory of Sampling.* New York: John Wiley and Sons.

Deming, W.E (1986) *Out of the Crisis.* Cambridge, MA: MIT.

Dempster, A.P. and Schatzoff, M. (1965) Expected Significance Level as a Sensitivity Index for Test Statistics. *Journal of the American Statistical Association*, **60**, 420–436.

Durant, W. (1953) *The Story of Philosophy*, 2nd ed. New York: Washington Square.

Dyson, F.J. (1988) *Infinite in All Directions.* New York: Harper and Row.

Eggleston, R., Sir (1978) *Evidence, Proof and Probability.* London: Weidenfeld and Nichol son.

Eisenhart, C. (1947) The Assumptions Underlying the Analysis of Variance. *Biometrics*, **3**, 1–21.

Efron, B. (2004) Statistics and the Rules of Science. *AMSTAT News*, **325**, 2–3.

Eisenhart, C. (1963) Realistic Evaluation of the Precision and Accuracy of Instrument Calibration System. *Journal of Research of the National Bureau of Standards*, **67C**, 21–47.

Eves, H. (1960) *Foundations and Fundamental Concepts of Mathematics*, 3rd ed. Boston: PWS-Kent.

Feather (1959) *The Physics of Mass, Length and Time.* Edinburgh: Edinburgh University Press.

Feynman, R.P.; Leighton, R.B.; Sands, M. (1975) *The Feynman Lectures on Physics.* Reading, MA: Addison-Wesley.

Fine, T.L. (1973) *Theories of Probability.* New York: Academic Press.

deFinetti, B. (1974) *Theory of Probability: A Critical Introductory Treatment,***vol 1**. New York: Wiley.

Fienberg, S.E. (1971) Randomization and Social Affairs: The 1970 Draft Lottery. *Science,* **171**, 255–261.

Fishburn, P.C. (1986) The Axioms of Subjective Probability (with Discussion). *Statistical Science*, **1**, 335–358.

Fisher, R.A. (1949) *The Design of Experiments*, 5th ed. New York: Hafner.

Fisher, R.A. (1956) *Statistical Methods and Scientific Inference.* New York: Hafner.

Frank, P. (1957) *Philosophy of Science.* Englewood Cliffs, NJ: Prentice-Hall.

Friedman F.; Pisani F.; Purvis F. (1978) *Statistics.* New York: Norton.

Frosch, R. (2001) Presidential Invited Address at JSM. *AMSTAT News*, **294**, 7–16.

Gastwirth, J.L. (1992) Statistical Reasoning in the Legal Setting. *The American Statistician,* **46**,55–69.

Giere, R.N. (1992) *Cognitive Models of Science.* Minneapolis, MN: University of Minnesota Press.

Ghosh, J.K., ed. (1988) *Statistical Information and Likelihood, A Collection of Critical Essays by Dr. D. Basu.* New York: Springer-Verlag.

Gnedenko, B.V. (1967) *Theory of Probability*. New York: Chelsea.

Hagood, M.J. (1941) *Statistics for Sociologists*. New York: Reynal and Hitchcock.

Hergenhahn, B.R. (1988) *An Introduction to Theories of Learning*, 3rd ed. Englewood Cliffs, NJ: Prentice-Hall.

Hempel, C.G. (1952) *Fundamentals of Concept Formation in Empirical Science*. Chicago: University of Chicago Press.

Hogben, L. (1957) *Statistical Theory*. New York: Norton.

Howson, C. and Urbach, P. (1989) *Scientific Reasoning the Bayesian Approach,* La, Salle, IL: Open Court.

Hull, D.L. (1990) *Science as a Process*. Chicago, IL: University of Chicago Press.

Imwinkelried, E. (1989) *Evidentiary Foundations*, 3rd ed. Charolottesville, VA: The Michie Company.

Inman, H.F.; Pearson, K.; Fisher, R.A. (1994) Fisher on Statistical Tests: A 1935 Exchange from Nature. *The American Statistician*, 48, 2–11.

Iranpur, R. and Chacon, P. (1988) *Basic Stochastic Processes THE MARK KAC LECTURES*. New York: Macmillan.

Irons, P. and Guitton S., eds. (1993) *May it Please the Court*. New York: The New Press.

Johnston, D.J. (1986) Tests of Significance in Theory and in Practice. *The Statistician*, **35**, 491–504.

Kadane, J.B. and Winkler, R.L. (1988) Separating Probability Elicitation from Utilities. *Journal of the American Statistical Association*, **83**, 357–363.

Kaye, D.H. (1987) The Admissibility of "Probability Evidence" in Criminal Trials—Part II. *Jurimetrics Journal* **27n2**, 160–172.

Kemeny, J.G.; Mirkil, H.; Snell, J.L.; Thompson, G.L. (1958) *Finite Mathematical Structures*. Englewood Cliffs, NJ: Prentice Hall.

Kempthorne, O. and Folks, L. (1971) *Probability, Statistics, and Data Analysis*. Ames: Iowa Press.

Kimble, G.A. (1961) Hilgard and Marquis' Conditioning and Learning, 2nd ed. Englewood Cliffs, NJ: Prentice-Hall.

Kline, M. (1980) *Mathematics: The Loss of Certainty*. Oxford: Oxford University Press.

Kolmogorov, A.N. (1950) *Foundations of the Theory of Probability*. New York: Chelsea.

Kuhn, T.S. (1962) *The Structure of Scientific Revolutions,* 2nd ed. Chicago, IL: University of Chicago Press.

Lehmann, E.L. (1986) *Testing Statistical Hypotheses*, 2nd ed. New York: John Wiley.

Lehman, E.L. (1993) The Fisher, Neymann–Pearson Theories of Testing Hypotheses: One Theory or Two? *Journal of the American Statistical Association,* **88**, 1242–1249.

Lindley, D.V. (1985) *Making Decisions*, 2nd ed. New York: Wiley.

Lindley, D.V., and Scott, W.F. (1984) *New Cambridge Statistical Tables*. Cambridge, U.K.: Cambridge University Press.

Loéve, M. (1955) *Probability Theory*. Princeton: Van Nostrand.

Lukasiewicz, J. (1929) *Elementary Logiki Matematycznej (Elements of Mathematical Logic)*. Warsaw: Panstomous Wydamnictwo Naukowe.

Maxwell, J.C. (1860) Illustration of the Dynamical Theory of Gases. In Niven, W.D., ed. *The Philosophical Magazine*, reprinted in *The Scientific Papers of James Clerk Maxwell*. New York: Dover Publications (1952).

McConway, K.J. (1981) Marginalization and Linear Opinion Pools. *Journal of the American Statistical Association*, **76**, 410–414.

vonMises, R. (1957) *Probability, Statistics and Truth,* 2nd ed.(English) London: George Allen and Unwin.

Morris, D.E. and Henkel, R.E., eds. (1970) The *significance Test Controversty*. Chicago: Aldine Publishing Company.

Morgan, J.R.; Chaganty, N.R.; Dahiya, R.C.; Doviak, M.J. (1991) Let's Make A Deal: The Player's Dilemma. *The American Statistician, 45*, 284–287.

Mueller, B. and Kirkpatrick, L. (1999) *Evidence, Practice Under the Rules*, 2nd ed. Gaithersburg: Aspen.

Nau, R.F. and McCardle, K.F. (1991) Arbitrage, Rationality and Equilibrium. *Theory and Decision, 31*, 199–240.

vonNeumann, J. and Morganstern, O. (1944) *Theory of Games and Economic Behavior*. New Jersey: Princeton University Press.

Neyman, J. (1950) *First Course in Probability and Statistics*. New York: Henry Holt and Company.

Ni, S. and Sun, D. (2003) Noninformative Priors and Frequentist Risks of Bayesian Estimators of Vector-autoregressive Models. *Journal of Economics, 115*, 159–197.

Nuclear Regulatory Commission (1975) Reactor Safety Study: An Assessment of Accident Risks in US Commercial Nuclear Power Plants. *NRC Report Wash.* 1400 (NUREG 751014), NTIS.

O'Hagen, A. (1994). *Kendall's Advanced Theory of Statistics, (vol. 2B) Bayesian Inference*. UK: Edward Arnold.

Persig, R.N. (1984) *Zen and the Art of Motorcycle Maintenance*. New York: Bantom Books.

Peto, R.; Pike, M.C.; Armitage, P.; Breslow, N.E.; Cox, D.R.; Howard, S.V.; Mantel, N.; McPherson, K.; Peto, J.; Smith, P.G. (1976) Design and Analysis of Randomized Clinical Trials Requiring Prolonged Observation of Each Patient, I: Introduction and Design, *British Journal of Cancer, 34*, 585–612.

Popper, K.R. (1983) *Realism and the Aim of Science*. London: Hutchinson.

Pratt, J.W. (1965). Bayesian Interpretation of Standard Inference Statements. (with Discussion). *Journal of the Royal Statistical Society, Series B, 27*, 169–203.

Rescher, N. (1969) *Many Valued Logic*. New York: McGraw-Hill.

Richards, R.J. (1987) *Darwin and the Emergence of Evolutionary Theories of Mind and Behavior*. Chicago, IL: University of Chicago Press.

Roberts, H.V. (1965) Probabilistic Prediction. *Journal of the American Statistical Association, 60*, 50–62.

Royall, R. (1986) The Effect of Sample Size on the Meaning of Significance Level. *The American Statistician, 40*, 313–315.

Royall, R. (1997) *Statistical Evidence, A Likelihood Paradigm*. London: Chapman & Hall.

Rubin, D.B. (1984) Bayesianly Justifiable and Relevant Frequency Calculations for the Applied Statistician. *Annals of Statistics, 12*, 1151–1172.

Ruhla, C. (1993) *The Physics of Chance*. Oxford: University Press.

Savage, L.J. (1962) *The Foundation of Statistics Inference* (A Discussion). London: Methuen.

Schervish, M.J.; Seidenfeld, T.; Kadane, J.B. (1990) State Dependent Utilities. *Journal of the American Statistical Association, 85*, 840–847.

Seidenfeld, T.; Kadane, J.B.; Schervish, M.J. (1989) On the Shared Preferences of Two Bayesian Decision Makers. *Journal of Philosophy, 86*, 225–244.

Senn, S. (2003) *Dicing with Death; Chance, Risk and Health*. Cambridge: Cambridge University Press.

Shewhart, W.A. (1931) *Economic Control of Quality of Manufactured Product*. New York: Van Nostrand.

Shewhart, W.A. (1939) *Statistical Methods from the Viewpoint of Quality Control.* Washington DC: Department of Agriculture.

Stone, M. (1960) The Role of Significance Testing: Some Data with a Message. *Biometrics,* **56**, 485–493.

Stone, M. (1961) The Opinion Pool. *Annals of Mathematical Statistics,* **32**, 1339–1342.

Thompson, B. (2006) A critique of *p*-values. *International Statistical Review,* **74**, 1–14.

Thompson, W. A. Jr. (1969) *Applied Probability.* New York: Holt, Rinehart and Winston.

Thompson, W.A. Jr. (1985) Optimal Significance Procedures for Simple Hypotheses. *Biometrika,* **72**, 230–232.

Todhunter, I. (1949) *A History of the Mathematical Theory of Probability.* New York: Chelsea.

Toulmin, S. (1963) *Foresight and Understanding, An Enquiry into the Aims of Science.* New York: Harper & Row.

Toulmin, S. (1972) *Human Understanding.* Princeton, NJ: Princeton University Press.

Tukey, J.W. (1960) Conclusions Vs. Decisions. *Technometrics,* **2**, 423–433.

Velleman, P.F. and Hogelin, D.C. (1992) Data Analysis. Chapter 2 of Hoaglin and Moore *Perspectives on Contemporary Statistics* (1992). Mathematical Association of America, Notes Number 21.

Varian, H.R. (1992) *Microeconomics Analysis*, 3rd ed. New York: Norton.

Wald, A. (1950) *Statistical Decision Functions.* New York: Wiley.

Weirahandi, S. and Zidek, J.V. (1981) Multi-Bayesian Statistical Decision Theory. *Journal of the Royal Statistical Society, Series A.*, **144**, 85–93.

White, H.E. (1958) *Continental Classroom.* NBC Television.

Wilder, R.L. (1983) *Introduction to the Foundations of Mathematics.* 2nd ed. Malabar, FL: Krieger.

Wilks, S.S. (1950) *Mathematical Statistics.* Princeton: Princeton University Press.

Wilson, E.B. (1952) *An Introduction to Scientific Research.* New York: McGraw-Hill.

Youden, W.J. (1962) *Experimentation and Measurement.* New York: Scholastic Book Services.

Author Index

Subject Index

Printed in the United States of America